INSIDE/OUTSIDE
OFFICE
DESIGN V 办公室设计V

深圳市艺力文化发展有限公司 编

华南理工大学出版社
SOUTH CHINA UNIVERSITY OF TECHNOLOGY PRESS
·广州·

图书在版编目（CIP）数据

办公室设计 = Inside/Outside Office Design.5：英汉对照 / 深圳市艺力文化发展有限公司编． — 广州：华南理工大学出版社，2015.8
 ISBN 978-7-5623-4647-0

Ⅰ．①办… Ⅱ．①深… Ⅲ．①办公室－室内装饰设计－作品集－世界－现代 Ⅳ．① TU243

中国版本图书馆CIP数据核字（2015）第112694号

办公室设计Ⅴ Inside/Outside Office Design Ⅴ
深圳市艺力文化发展有限公司　编

出 版 人：韩中伟
出版发行：华南理工大学出版社
　　　　　　（广州五山华南理工大学17号楼，邮编510640）
　　　　　　http://www.scutpress.com.cn　E-mail: scutc13@scut.edu.cn
　　　　　　营销部电话：020-87113487　87111048（传真）
策划编辑：赖淑华
责任编辑：陈　昊　黄丽谊
印 刷 者：深圳市汇亿丰印刷科技有限公司
开　　本：595mm×1020mm　1/16　**印张**：26.5
成品尺寸：248mm×290mm
版　　次：2015年8月第1版　2015年8月第1次印刷
定　　价：448.00元

　　　　　版权所有　盗版必究　　印装差错　负责调换

全球优质设计作品，
尽在 ACS 创意空间

―― **WWW.ACS.CN** ――

最新锐的创意灵感　最前沿的设计概念

《ACS 创意空间》―― 国际青年设计师协会官方指定合作媒体

关注有惊喜！

扫描二维码，开启电子阅读体验，海量优秀作品随心看！

ARTPOWER　ACS

Preface 序言

Many tempers come together in one area called "office". Dozens if not hundreds of people with different ambitions and desires come here, and one of the main tasks an architect and a client work on is to do their best to ensure that a job will be done in a comfortable environment. The more convenient the working space, the better the time employees have there, the better the output of their work is.

As architects we find it very appealing that companies go for individuality. Now every project has become an opportunity to create a one-of-a-kind space that reflects a company's business, becomes its brand identity and the best possible working place for a team. More and more companies recognize that apart from a workplace, meeting rooms and utility rooms their staff needs places for so called random meetings. Places where an interesting idea may come up, and where impromptu brainstorms will take place. Informal setting of such places allows one to unwind and catch a break and then proceed to work with renewed vigour. With increasing frequency we create offices with game zones, porches, several coffee-points, and small spaces for phone conversations or meetings, where employees can contemplate in peace and unwind or vice versa – discuss things on the fly. It is more common now that we start out from the needs of teams, and we see more often that businesses care for their staff as much as for their clients. This trend allows creation of interesting creative spaces, unlike unchanging offices of the past. We are curious to know what the years to come will bring, but for some of our clients the future has already come today!

Amir Idiatulin,
CEO IND Architects, IND Office

　　性情各异之人聚集在一处，名为"办公室"，这是千百位抱负与追求不同之人相聚的地方，因而建筑师和客户的主要任务是全力以赴，打造舒适宜人的工作环境。往往办公空间愈舒适，职员会拥有愈快乐的工作时光和愈高的工作效率。

　　企业越来越追求个性化，这何尝不引起建筑师的关注。如今，每个项目都意味着一个创造独一无二空间的机会，来体现客户的公司业务、树立品牌形象，为其团队打造最好的办公空间。越来越多的公司也认识到，仅拥有办公室、会议空间及储存空间是远远不够的，职员也需要自由会谈空间，这正是有趣的想法诞生和即兴头脑风暴进行的地方。这类空间中非正式的布置有利于人们放松和休息，抖擞精神继续工作。越来越多的办公室中加入了游戏区、门廊、咖啡区以及其他通话空间和小型会议空间。此外，还有可供职员静心思考、休息或进行激烈讨论的地方。时至今日，从团队的需求出发来开展设计工作也越来越常见。我们也发现，企业对职员的关注已不亚于对客户的关注。在这种趋势之下，打造更多富有创意的趣味空间已成为可能，它们与以往一成不变的办公室完全不同。我们好奇未来会发生什么，但对有些客户来讲，未来已经成为现在。

Amir Idiatulin,
CEO IND Architects, IND Office

CONTENTS 目录

002 | ADVERTISING/MARKETING

- 002 Velti Headquarters
  Velti 总部
- 008 Innocean Headquarters Europe
  Innocean 欧洲总部
- 016 BBDO Group Advertising Agency Moscow Representative Office
  BBDO 广告集团莫斯科代表处

024 | ARCHITECTURE FIRM

- 024 Astarta's Moscow Office
  阿斯塔尔塔莫斯科办公室
- 030 Office Design of IND Architects
  IND Architects 工作室

038 | FINANCIAL/BUSINESS SERVICES

- 038 HSB
  HSB 办公空间
- 050 Alfa Bank Office
  阿尔法银行办公室
- 062 The Gores Group Headquarters
  格雷斯集团总部
- 074 Co-Work Angel Office
  Co-Work Angel 办公室
- 082 Algomi
  Algomi 办公空间
- 090 ZS Associates
  致盛咨询公司办公室

102 | FOOD/BEVERAGE

- 102 Red Bull - Cape Town Headquarters
  红牛开普敦总部
- 112 Head Offices, Red Bull Canada
  加拿大红牛总部
- 120 Coca-Cola London Headquarters
  可口可乐伦敦总部

126 | FURNITURE / PRODUCT

- 126 One Workplace Headquarters
  One Workplace 总部
- 138 PETZL North American Headquarters and Distribution Center
  PETZL 北美总部和配销中心
- 146 American Standard
  美标办公室

156 | GAMING

- 156 Chartboost San Francisco Office
  Chartboost 旧金山办公室
- 162 Lego Turkey
  土耳其乐高办公空间

166 | HARDWARE / SOFTWARE DEVELOPMENT

166	Microsoft Gurgaon	
	吉尔冈微软办公室	
182	Microsoft - Redmond Building 44 Offices	
	微软公司——雷德蒙 44 号办公楼	
190	Meltwater Tenant Improvement	
	融文集团办公室	
196	Autodesk Israel - Tel Aviv	
	以色列特拉维夫市欧特克开发中心	
204	Headquarter of Microsoft in Lisbon	
	里斯本微软总部	

216 | MANUFACTURING

216	Unilever Brand Hub Europe	
	联合利华欧洲品牌中心	
226	NUON Amsterdam	
	阿姆斯特丹 NUON 总部	
236	Unilever Algida Ice Cream Factory	
	联合利华 Algida 冰激凌工厂	

242 | MEDIA /PUBLISHING

242	BBC Worldwide Americas	
	英国广播公司美洲总部	
250	MTV Networks Headquarters	
	音乐电视网总部	
258	Wunderman/Bienalto Sydney	
	悉尼 Wunderman 和 Bienalto 公司办公室	

266 | MISCELLANEOUS

266	'T PARK	
	T 公园办公空间	
272	Integral Iluminación Comercial Building	
	Integral Iluminación 商业大楼	
280	EMKE Office Building	
	EMKE 办公大楼	
286	Mi9 Sydney	
	Mi9 悉尼办公室	
296	Uniform	
	Uniform 办公空间	
304	Grupo CP Corporate Interior	
	Grupo CP 公司办公室	
320	Silos 13	
	13 号筒仓	
328	Talent Garden office, Brescia - Italy	
	意大利布雷西亚 Talent Garden 办公室	
336	Johnson & Johnson Eyesight Health Institute	
	强生视力健康研究所	
344	EB Group Showroom & Office	
	EB 集团展厅和办公室	

352 | NON-PROFIT/GOVERNMENT

352	Headquarters of the Fondation Jérôme Pathé Seydoux	
	百代基金会总部	
362	NS Stations	
	NS Stations 办公室	
372	Northern Territory	
	泰利驿站	

380 | SOCIAL/WEB/ONLINE

380	Google Amsterdam	
	阿姆斯特丹谷歌办公室	
386	Walmart Headquarter in São Paulo	
	圣保罗沃尔玛总部	
400	Tencent Guangzhou Office	
	腾讯广州办公园区	

410	Contributors	
	设计师名录	

INSIDE

›OUTSIDE

ADVERTISING/MARKETING

Velti Headquarters

Velti 总部

Architects and Interior Design Agency: AECOM

Architecture Project Team: Michelle Ives, Associate IIDA, LEED AP – Firm Principal / Partner in Charge, Allard Kuijken, LEEP AP – Associate Principal, Kelly Capp, Associate IIDA – Senior Associate

Interior Design Project Team: Michelle Ives, Associate IIDA, LEED AP – Firm Principal / Partner in Charge, Allard Kuijken, LEEP AP – Associate Principal, Kelly Capp, Associate IIDA – Senior Associate

Client: Velti mGage

Location: San Francisco, CA, USA

Area: 3,530 m²

Photography: David Wakely Photography

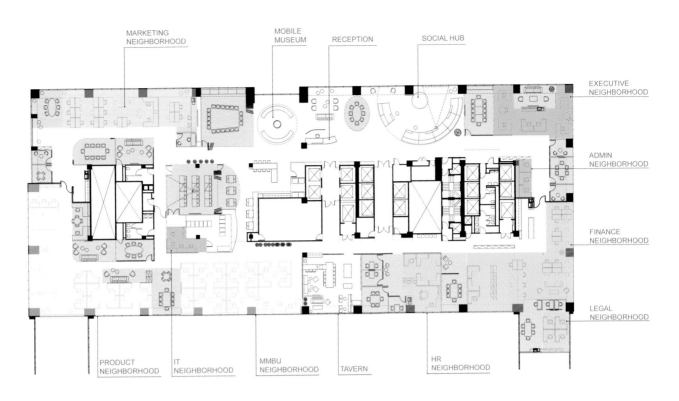

FLOOR PLAN

- MEETING SPACE
- SOCIAL SPACE
- WORK SPACE

- MARKETING NEIGHBORHOOD
- PRODUCT NEIGHBORHOOD
- FINANCE NEIGHBORHOOD
- EXECUTIVE NEIGHBORHOOD

Velti, a leading global mobile marketing and advertising firm, wanted their new headquarters to redefine the dynamics of the modern workplace. This meant creating a workspace that gives its employees flexibility in how they work. Located on the sixth floor of the Steuart Tower in San Francisco's Financial District, this new 3,530 m² headquarters offers expansive views of the Ferry Building and San Francisco Bay and is filled with common areas designed to accommodate a variety of interactive working situations. The look and feel of the workspace supports Velti's 200 employees and their broad range of preferred work styles. The program includes offices, dining facilities, exhibition and public gathering space as well as an employee recreation lounge – the Tavern – which houses a video arcade, photo booth, bar, and game room. Small tables for intimate conversation, large tables for large formal meetings, stadium seating for company meetings, and moving partitions all create a workplace that encourages individuals to work in a manner to best suit their needs. With carefully attuned circulation, retractable walls, ample natural light, formal and informal workspaces – the office has a sense of being a physical social network. Meeting, social and work spaces are laid out in a clever departmental neighborhood system that allows marketers, engineers, executive directors, IT, product developers, and in-house counsel to work separately or independently as desired. A clear aesthetic language defined by consistent color, graphics, furniture, and finishes, separate each department and provide a sense of individuality for each respective group. All the neighborhoods are linked by exposed concrete floors, open and layered ceilings, custom super graphics, and predominant textures and graphics that run throughout the headquarters and set the tone for a workplace that values collaborative creativity.

Velti 是全球领先的移动营销和广告公司。该公司希望新总部能重新诠释现代办公空间动力学，这就意味着为员工创造一个灵活的办公空间。新总部设在旧金山金融区斯图尔特大厦的第 6 层，面积约为 3 530 m²，其位置优越，视野开阔，能观赏到轮渡大厦和旧金山海湾的风光。空间内包含了各种公共空间，适合各种互动工作场合。办公空间的外观和触感不仅迎合了 Velti 公司的 200 名职员的品味，还能适应各种各样的工作模式。该项目涉及办公室、餐厅设施、展厅、公众聚集空间和职员休闲区（也叫做酒馆，内设电子游戏室、照相亭、酒吧和游艺室）。空间内摆放的小桌子适合亲密的交谈；大桌子适合大型正式会议；露天阶梯空间适合举行公司会议；移动分区则最适合单独工作。精心规划的流通线路、可伸缩墙、充足的自然光线，以及各种正式和非正式工作空间，使得办公室成为了一个有形的社交网络。会议、社交和工作空间的布局十分巧妙，就像一个分部的领域体系。营销人员、工程师、执行董事、IT 人员、产品开发人员和内部法律顾问都可以在这个体系中单独工作。色彩、图案、家具和墙饰组成了清晰的美学语言，将每个部分分隔开来，赋予每个团队个性。而裸露的混凝土地面、多层次开放天花板、专门打造的超大图案以及醒目的纹理和图形，贯穿了整个总部，并将各个部门连接起来，为这个重视协作和创新的办公空间奠定了基调。

ADVERTISING/MARKETING

Innocean Headquarters Europe

Innocean 欧洲总部

Design Agency: Ippolito Fleitz Group GmbH

Project Team: Andrew Bardzik, Anke Stern, David Schwarz, Frank Peisert, Sebastian Tiedemann, Yuliya Lytyuk, Gunter Fleitz, Peter Ippolito, Daniela Schröder, Tim Lessmann

Client: Innocean Worldwide Europe GmbH

Location: Frankfurt, Germany

Area: 2,800 m²

Photography: Robert Hoernig

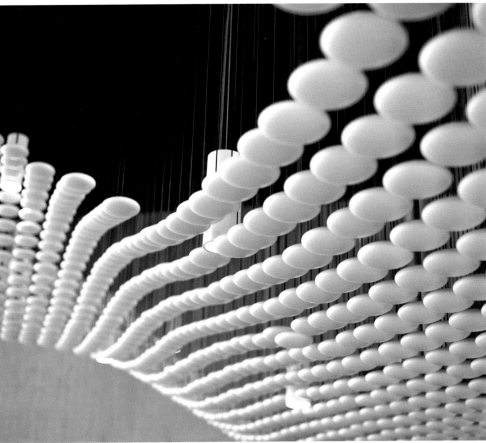

The internationally operating advertising agency Innocean with headquarters in Korea has moved into new European headquarters in Frankfurt am Main. A flexible and modern work world was created for the young, design-conscious company, which fits the different work zones within the agency.

Dynamism and movement are key features of the design, which assimilates employees and visitors the second they enter the spacious reception hall. These design elements guide you through the open work zones and the specially created employee library right up to the in-house gym on the fifth floor, which offers an amazing view over Frankfurt. Polygonal spatial elements and a wide range of materials represent the high design standards of the agency itself. Open and transparent work areas, paired with semi-public and completely discreet conference zones promote a creative and communicative working atmosphere.

国际经营广告公司 Innocean 将原位于韩国的总部转移到欧洲，在法兰克福经济中心设立了新的总部。这一灵活的现代工作空间是专门为这家年轻并且注重设计的公司设计的，各种分区适合公司内的不同工作部门。

活力和运动是设计的关键因素，当职员和访客进入宽敞的接待大厅时，就会被深深地吸引住。这些设计元素引导着人们在开放式的工作空间内移动，从设计独特的职员图书室可以直接进入 5 楼的室内健身房，在健身房中则能够欣赏到迷人的法兰克福城市风景。多边形空间元素和各种各样的材料体现了设计公司的设计水准。开放、透明的工作空间，与半开放、全封闭的会议空间，共同营造出可促进创新和互动的工作氛围。

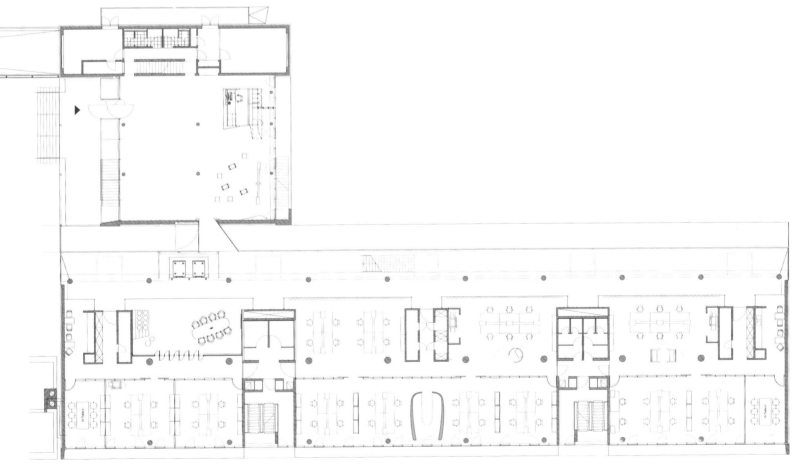

ADVERTISING/MARKETING

BBDO Group Advertising Agency Moscow Representative Office

BBDO 广告集团莫斯科代表处

Architects: VOX Architect

Chief Architect: Boris Voskoboinikov, Maria Akhremenkova (interior designer), Dmitry Ovcharov, Maxim Frolov (3D)

Chief Engineer: Sergey Kurepin

Project Group: Margarita Kornienko, Viktor Kolupaev, Maria Nasonova, Olga Ivlieva

Client: BBDO GROUP, Moscow

Location: Moscow, Russia

Area: 3,400 m²

The main challenge that faced Nefaresearch's architects, was to reconstruct four-storey factory building of 19th century to the office for advertising agency. The style of the office had to resemble an art center, rather than traditional office. The aim was to create with irony a rich dynamic and contemporary image in the BBDO's branded palette (red, white, grey and black), covering and unifying all office spaces and in the same time dividing them into functional zones. The image should also contribute to creative and non-standard thinking.

Chief Architect and the founder of the studio – Boris Voskoboinikov suggested to define ground floor as a showcase, demonstration the image of the company's from the street and to make the front group a powerful style-forming center. He developed the concept that perfectly reflecting the nature of BBDO Group.

The project covers the whole four-storey building. The ground floor is a comfortable common area, open to guests and clients; the remaining three floors, on the contrary, destined to be agency's working space. The reception at the entrance separates people flows: company's employees, heading for working area, and guests and clients, who came for a meeting or to the cafe.

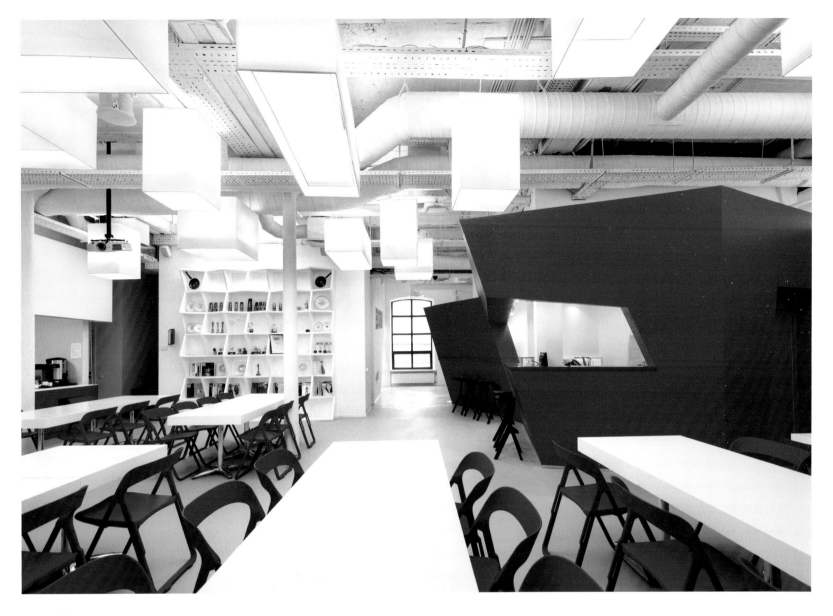

1 floor

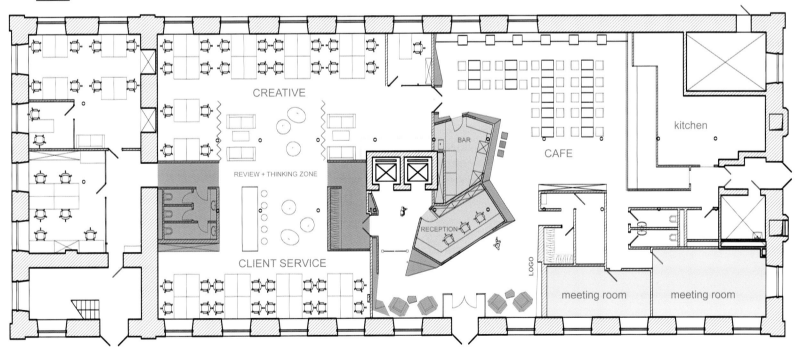

The cafe space is easily transformed and has several additional functions, such as carrying out seminars, discussions and design reviews conduction, as well as Digital Lab technology testing. In other words, this space is a public center of the agency.

The red structure in the most dynamic point acts as the reception and the bar simultaneously, as if growing out of the walls of the building. Developing in the space, gaining strength, focusing on attention and becoming whole room, it weakens and disappears completely, remaining as a slight accent. This architectural solution of the plan made it possible to cope with the impossibility of transferring many partitions and engineering communications. The red structure, designed by Boris Voskoboinikov and Maria Akhremenkova (the interior designer of the project) is a kind of navigation for the visitors to the office.

Workspaces are interesting for its planning solution. This decision is dictated by a typical problem of advertising agencies: the constant process of changing of the structure, changing number of employees and departments, as well as modification of the interrelationships of these departments.

Advertising agency is a sensitive to the external situation body and, respectively to this, the workspace should be reconfigurable to the new conditions instantly. The administrative core, located in the center of each working floor, implements the functions of the offices and conference rooms. Depending on a necessity, the offices can be transformed into a meeting room and vice versa.

For its design the administrative core embodies the idea of "Architecture in the interior". "Interior houses" organize the space in playful dynamism. Around these "houses" the mobile office space is formed along the windows, and its sections are separated by soft relaxation zones. Thanks to this structure, it is possible to easily reorganize both the entire floor and working areas without any major effort.

The height of the ceiling on the top floor allowed making a mezzanine for a relaxation or a comfortable solitary work.

Nefaresearch 的建筑师所面临的主要挑战为将一栋19世纪的4层厂房改造成BBDO集团的办公室。办公室的风格不能是传统的办公室，而应该像一个艺术中心。项目旨在用反讽的设计手法创造一个充满活力的现代形象，用BBDO的品牌色彩（红、白、灰、黑）覆盖所有的办公空间，在将所有空间连为一体的同时，也将它们划分成不同的功能区。这种形象还能激发创新的思想和奇异的想法。

首席设计师和集团创始人Boris Voskoboinikov建议将一楼设计成引领潮流的展示中心，并从面向街道处就开始展示公司的形象。这种理念完美地体现了BBDO集团的行业属性。

项目包含了4层楼的设计。1层是舒适的公共空间，向来宾和客户开放；而其余3层则是集团的工作空间。流动的人员在入口处的接待区开始分流：公司职员走向工作区；来宾和客户走向会议室或咖啡区。咖啡区的设置十分灵活，改造起来十分轻松。另外，它还有其他几项功能，例如，在此还能够进行研讨、设计评审公告以及数字实验室技术测试。换言之，这里是集团的公共活动中心。

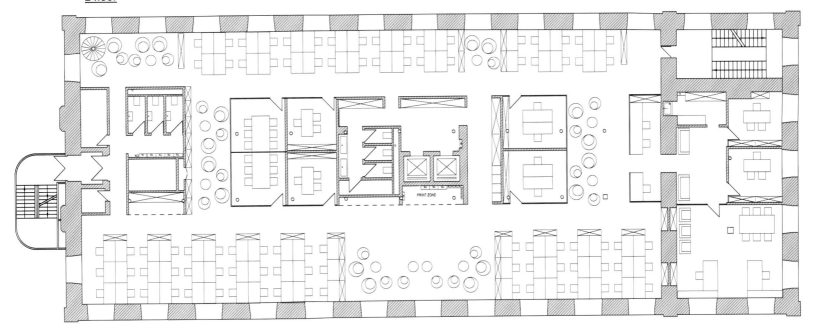

2 floor

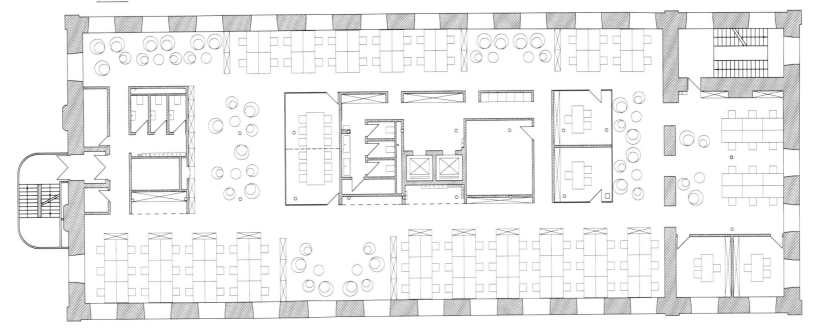

3 floor

接待台设在空间内流动性最强的地方，这个红色结构就像是从墙上生长出来的，还充当了吧台。它在空间内成形，获取力量，进而成为了焦点，接下来又延伸至空间的各个角落，然后慢慢变弱，最终消失不见，仅有一丝余音保留下来。这种建筑布局策略让分区转型和工程通信都成为了可能。由Boris Voskoboinikov和Maria Akhremenkova（负责该项目的室内设计师）设计的这个红色结构，对来访的客人具有导航作用。

这个工作空间因其布局策略而变得十分有趣。这一决定解决了广告公司的典型问题：结构、员工人数、部门数量以及部门之间的关系都在不断地发生改变。

广告公司对外部形式十分敏感，考虑到这一点，工作空间需要根据新形式不断地重新改造。位于每个工作楼层中央的行政中心设有办公室的各功能区和会议室。若有需要，这些空间可在办公空间和会议室之间相互转换。

行政中心蕴含了"楼中楼"和"房中房"的理念，使空间更有活力和趣味性。这些"房屋"周围是沿着窗户而建的一栋办公区，而且各种分区之间设有舒适的休闲区，将它们分隔开来。这种布局使得整层楼和各个工作区的重新布置成为了可能，而且十分便捷。

顶楼天花板较高，因此还设计了中间夹层。夹层上设有休闲区和舒适的个人工作空间。

4 floor

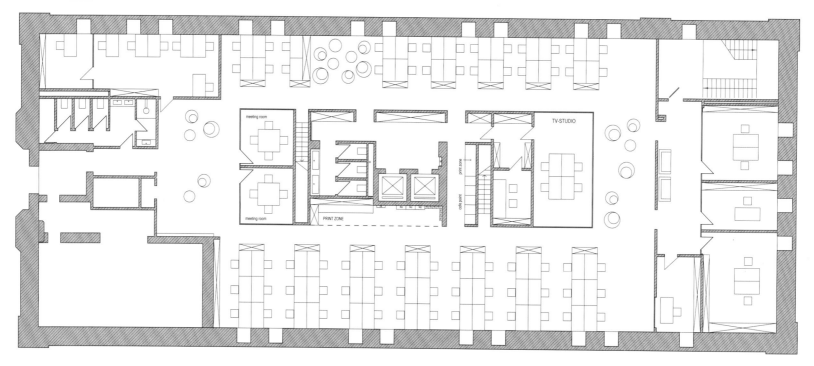

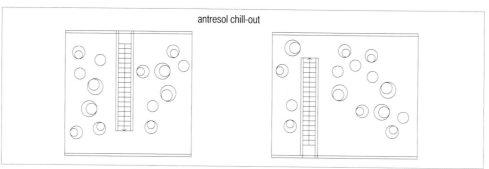

antresol chill-out

ARCHITECTURE FIRM

Astarta's Moscow Office

阿斯塔尔塔莫斯科办公室

Designers: Tatyana Romanova, Galina Bandurkina, Ilya Starostin, Tatyana Tsimbalistova

Client & Contributor: Astarta

Astarta company is located in Moscow, Russia. The updated office of it was started in July, 2013.

The main task of new office was creation of the atmosphere in which partners, employees and Astarta's friends, coming to office, saw not only habitual studies, and every day opened for itself something new and interesting. We wanted to create not simply a hall and a show-room, but different variations of design ceilings, walls and furniture in uniform and operating office interior.

Each premise of office: expectation zones, negotiation offices, lounge-zones, open spaces are issued in different style. However, all rooms are integrated by one general idea – a wave. The wave is a symbol of changes, dynamics and development. The wave is shown in soft, smooth bends of details of the interior, slightly rounded-off geometry of furniture and lamps.

The main materials, used at creation of office, are milled MDF and gypsum cardboard. All materials are most technological and safe. Besides, it is possible very quickly and to mount them easily.

The office area quite big is 504 m². "We wanted to create additional workplaces for growing group of

employees. All rooms are issued in such a way that here it is possible not only to work, but also to hold presentations, to give lectures and to invite more and more clients," – Galina Bandurkina, founder of Astarta explains.

Primary color of walls at office is neutral yellow which doesn't hurt the eyes of employees during the work. Ceilings white, however lighting can change color and create in an interior the special atmosphere, depending on the necessary situation – presentation, a relax or demonstration to clients of production of the company.

Designers first of all took care of functional zoning of the room. Partitions are executed in neutral yellow color that employees don't distract from work. White neutral ceilings thanks to illumination can change the color, creating the necessary atmosphere indoors.

The brightest detail of an interior is the central partition which divides office into entrance and working zones. Plot of the photo printing put on glass, macro shooting of elements of the nature – flowers, fruit, water drops acts. This interesting find says that the world around is much brighter, than it seems.

Lounge-zones are executed in bright, contrast colors. Walls with a photo printing put on glass help to create effect of a relaxation.

阿斯塔尔塔公司位于俄罗斯莫斯科。2013 年 7 月，其新办公室开始投入建设。

该项目的主要任务是营造一种氛围，当员工、合作伙伴和公司的友人进入办公室时，不仅能看到常见的东西，每天还能获取新鲜有趣的事物。我们想创造的不仅仅是一个大厅和一个展厅，还融入了多样的天花板设计和一致的墙壁、家具设计。

办公室设有接待区、洽谈区和休息区，各种公共空间的风格各不相同。然而，一种叫做"波浪"的核心思想却贯穿所有的空间。波浪象征着变化、动力和发展，它不仅体现在室内柔软和流畅的弯曲细节中，还体现在偏圆的几何形家具和灯具中。

办公室装饰是以中密度纤维板和石膏纸板为主要材料。所有的材料都是采用高技术做成，安全性高，安装起来也十分简单快捷。

办公区十分宽敞，占地 504 m²。"因为员工数量在不断地增长，因此，我们想预设额外的工作空间。构想的所有空间是不仅能够工作，还能进行展示、举办讲座以及邀请众多客户，"阿斯塔尔塔公司创办人 Galina Bandurkina 说道。

办公室的主要色彩为淡黄色，当员工在工作时，不会刺激眼睛。天花板采用了白色，而灯光可以根据场合转变色彩，比如在展示、放松和为客户演示产品等场合时，可以营造相应的氛围。

设计师首先关注的是功能空间。各分区也采用了淡黄色，不会影响员工工作。白色天花板在灯光的照耀下可以改变色彩，能根据需要营造出适当的室内氛围。

室内最显眼的细节是中央的分区，该分区将空间划分成了入口区和工作区。在这一空间中，玻璃上的照片被放大，花朵、水果和水滴灯等自然元素都被清晰地展示出来，以一种有趣的方式说明其实际上更加明亮。

休息室采用了明亮的对比色彩。玻璃墙上贴着的照片，使空间氛围更加轻松。

ARCHITECTURE FIRM

Office Design of IND Architects

IND Architects 工作室

Design Agency: IND Architects

Location: Moscow, Russia

Area: 280 m²

Photography: Alexey Zarodov

When working on the project of IND Architects office, we entrusted ourselves with the task to create a contemporary creative space with a minimum of distracting elements, so that designers and architects could focus on their projects as much as possible. The office has become a business card and a reflection of the studio which has implemented a number of projects over 5 years of its existence.

The office has been arranged in ARTPLAY Design Center and has been decorated in the loft style. Designers have shown the benefits of a former industrial premise to their best advantage – they have kept the double-floor height area in some places, have deliberately left the concrete ceiling panels unpainted and have retained the original concrete structure like timber shutter texture on the first floor and ceiling panels on the second floor. Exposed black-colored utilities accentuate the industrial past of the building and form a contrast to white walls of the office.

The office has been divided into two zones – one is a volumetric double-floor height area for designers and another is a cozy space on the second floor for architects. The walls of the latter one are used for attaching the pictures and drawings of studio's current projects. In addition to two open space zones, a reception zone, a meeting room, and a coffee-point have been arranged on the first floor; an office, a leisure area, and an open meeting room for the staff may be found on the second floor. 2 or 3 person meetings may also be held at a round table in the first-floor open space. 2 separate wardrobes have been arranged for the staff and guests - one is at the entrance and another is at the meeting room.

Grey and white are basic interior colors. Bright yellow details – such as infographics, a creatively different full-wall unicorn in the first-floor open space, and small items, like flower cache-pots, desk folders, and décor details stand out sharply against quiet shades. The infographics had been developed as a part of a new corporate style of IND Architects and was patterned by the studio's architects.

In the leisure area, the employees may catch a break in a work process and talk about their goings, interesting projects and news while playing darts or foosball or read an interesting book nestling themselves down on a convenient pouf.

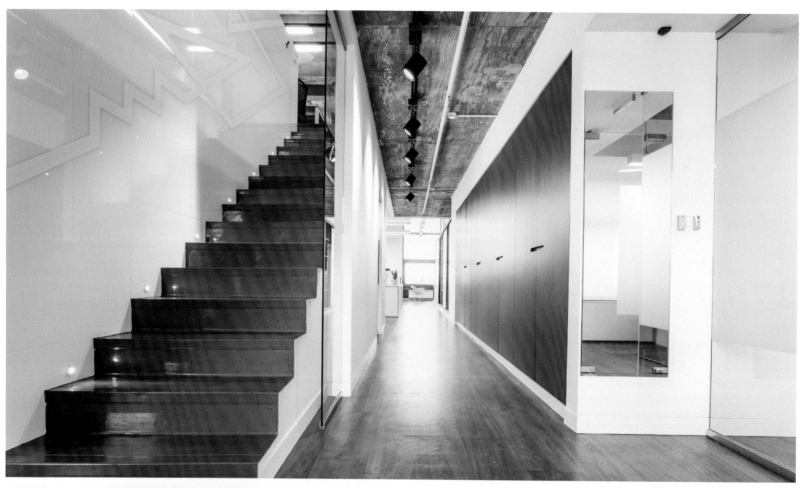

　　该项目旨在尽量避免使用分散注意力的元素而打造出一个现代化的创意空间，从而能让设计师和建筑师专心致志地完成自己的项目。IND Architects 工作室就像一张名片，展示了工作室自成立以来 5 年间设计的作品。

　　办公室位于 ARTPLAY 设计中心，装饰方面则采用了 LOFT 的风格。设计师通过保留部分 2 倍高空间和原混凝土结构，维持原建筑的混凝土天花板的模样，一楼的木窗和二楼的天花板也都是保留下来的，充分展示了原工业基地的优势。暴露的黑色建筑设施增强了建筑的工业风格，与办公室的白墙形成鲜明的对比。

　　办公室分为 2 个区域：专为设计师打造的两倍高的宽敞空间，以及为建筑师打造的楼上舒适空间。二楼空间的墙壁上展示了工作室的照片、设计图和现有的项目。除了这 2 个开放的空间以外，一层还设有接待区、会议室和休息室；而员工办公室、休闲区和开放式会议室都设在二楼。一层的开放式空间中设有圆桌，2 到 3 人的会议可以在此进行。此外，还为员工和客户专门设了 2 个独立的橱柜，一个设在入口处，另一个设在会议室中。

　　室内主要以灰色和白色为主色调，还掺入了亮黄色细节，比如信息图表、一楼公共空间墙壁上的创意独角兽，以及诸如花盆、文件夹和饰品等小物品。这些细节在冷色调背景的衬托下显得尤为突出。信息图是 IND Architects 的新合作方式，是由工作室的建筑师制成。

　　员工在工作过程中可以到休闲区小憩一会，闲谈近况，分享有趣、新鲜的项目，或者是玩飞镖和桌上足球，或是坐在舒适的椅子上阅读有趣的书籍。

FINANCIAL/BUSINESS SERVICES

HSB

HSB 办公空间

Interior Architecs: pS Arkitektur

Head Architect: Peter Sahlin

Project Architect: Beata Denton

Assisting Architects: Martina Eliasson, Thérèse Svalling, Emilie Westergaard Folkersen

Location: Stockholm, Sweden

Area: 9,000 m²

Phototgraphy: Jason Strong Photography

HSB is Sweden´s largest housing developer and owned by its members. It´s Stockholm office has just undergone a complete renovation in order to allow for openness and accessibility. All of this applying to the environmental classification called "Miljöbyggnad Silver". pS has been the interior architect and space planner. Much effort has gone into creating an ergonomic and modern office in terms of acoustics and lighting. This in combination with new technology has made the change from cellular offices to open workspace a pleasant experience.

Social interaction and energy has been the keyword and the theme is "Welcome home"! The reception, the so called "living shop" and the inner courtyard all merge together on the ground floor, allowing for staff and guests to mix and mingle informally. The interior design is comfortable and colourful, contrasting efficiently against the original 12.2 m intarsia wall and paternoster lifts.

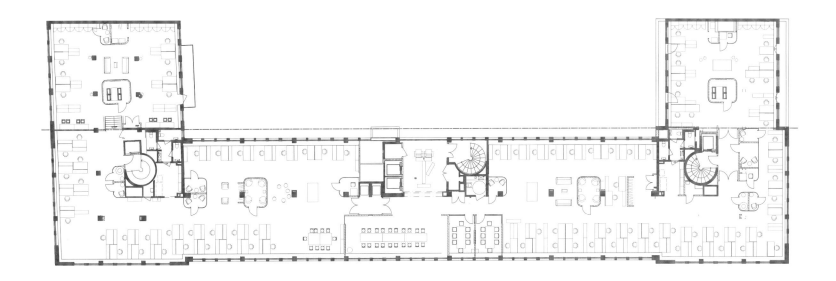

Plan 6
1:400

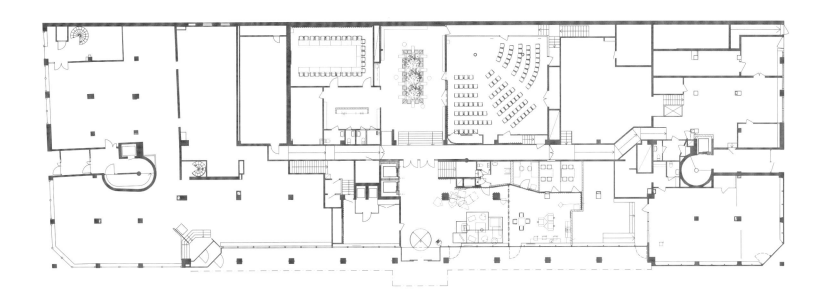

Plan 1
1:400

The office space takes its inspiration from the city block. Each block consists of a number of desks and in the centre there are "squares" and meeting points such as lounge furniture, hotdesks and telephone booths.

About 420 people work in the building. The top floor has an amazing view over the city roof tops and presents a dozen or meeting rooms for external meetings. Relaxing lounges and a creative space named "Think Tank" completes this welcoming office.

　　HSB 是瑞士最大的地产开发企业，公司位于斯德哥尔摩的办公室刚刚完成了翻修，空间变得更加开敞和通畅。空间的分类遵循了可持续发展评级体系银牌认证标准——Miljöbyggnad Silver。pS 建筑公司是该办公空间的室内建筑师和空间规划师。不管是声音效果，还是灯光效果，或者是人造环境和现代空间的创造，他们都投入了很多的精力。新科技的应用更是让移动办公转变为开放办公的体验令人更加愉悦。

　　社交和能源是设计的关键词，而主题则是"欢迎回家"。一楼的接待区，也就是所谓的"生活店"，与内部的庭院连为一体，给职员和访客提供了交际空间。空间的内部设计舒适无比，色彩丰富，与12.2 m 的原木板墙、链斗式电梯形成鲜明的对比。

　　办公空间的灵感来自城市的街区，空间内有许多区，每个区布置了许多椅子，中央则设了躺椅、公用办公桌和打电话的小房间，使得这儿成为可以聚集讨论或谈天的"小广场"。

　　公司约有 420 位职员，办公室足以容纳所有职员。顶层能远眺美丽的城市风景，针对外部会议设有十几间会议室。休息空间和"智库"创意空间让这个倍受员工喜爱的办公室更加完美。

FINANCIAL/BUSINESS SERVICES

Alfa Bank Office

阿尔法银行办公室

Design Agency: IND Architects

Location: Moscow, Russia

Area: 2,630 m²

Office Construction: RD Construction

Photography: Andrey Jitkov

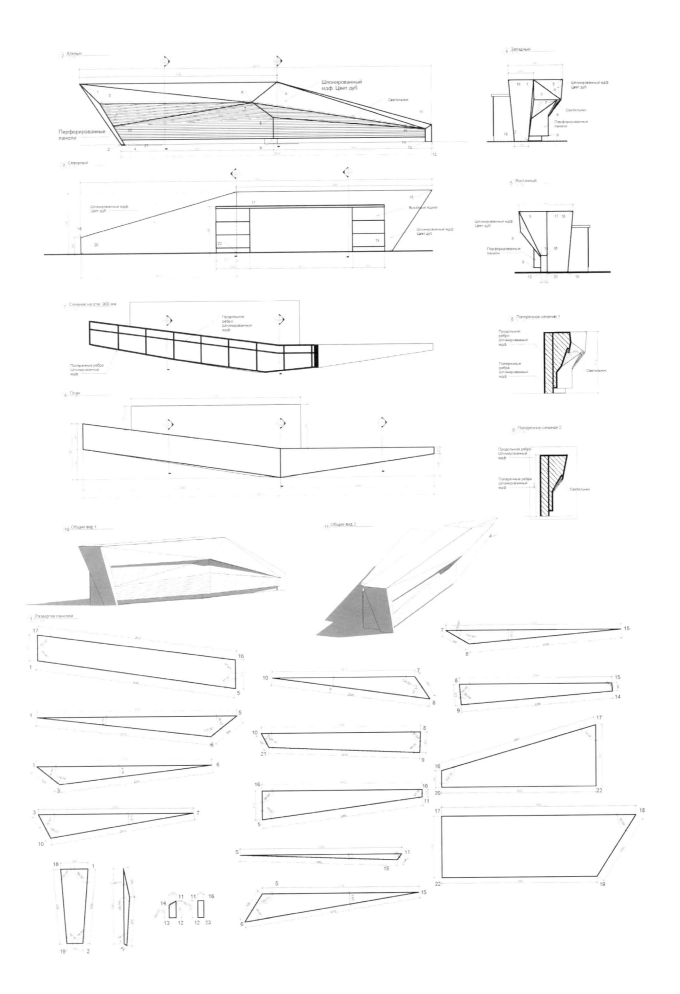

One of the unique features of the office designed by architects of IND Architects for one of Alfa Bank's branches is, first of all, its creativity which is incredible for the financial sector. Functional, striking, and innovative, it will be a place to work in for young and vigorous employees of Alfa Laboratory – a special unit of Alfa Bank engaged in electronic business.

The idea of the interior is based on superheroes and street art – the components which Laboratory employees can relate themselves to. The finishing materials used are as follows: textured concrete, wood, perforated metal, various kinds of glass – clear, dim, and patterned. A bright carpet tile facilitates the navigation – meeting zones are colored, while various circles in an open space zone help employees to find required groups and departments.

Walls of the office combine several functions: a decorative function – bright wall murals featuring superheroes and comic books; a practical function – a special surface where you can write with felt-tip pens; a bulletin board material has been pasted on some walls, where employees can fix their materials; an informative and motivational function – a wall with quotes of great people. Ceilings in a presentation and game zones are decorated with human silhouettes, which stands for a team spirit. There are lamps between the silhouettes – they represent creative thoughts and fresh ideas. Besides, such solution improves the room acoustics.

A distinguishing feature of the Laboratory is that employees can not only work on their work places, but take an active part in brainstorms and meetings as well. This feature has identified the laying out of the office – there are many meeting zones and buzz session zones (coffee points); the game zone to have a rest with a ping-pong table, various board games, and carpet-covered walls to play darts; and two outdoor porches with gleamy furniture here. There is an open meeting hall in the center of the office – a place to make presentations with a gong on the wall. There are LCD screens in front of meeting rooms which display if the rooms are available or not.

Various lighting solutions have been implemented in the office – linear light in the open space zone, LED backlighting in a hallway, and soffits in the game and the presentation zones. The classy, dynamic, and functional design with striking elements and interesting details – this is not merely an office, but a really comfortable place to create unusual and contemporary solutions too.

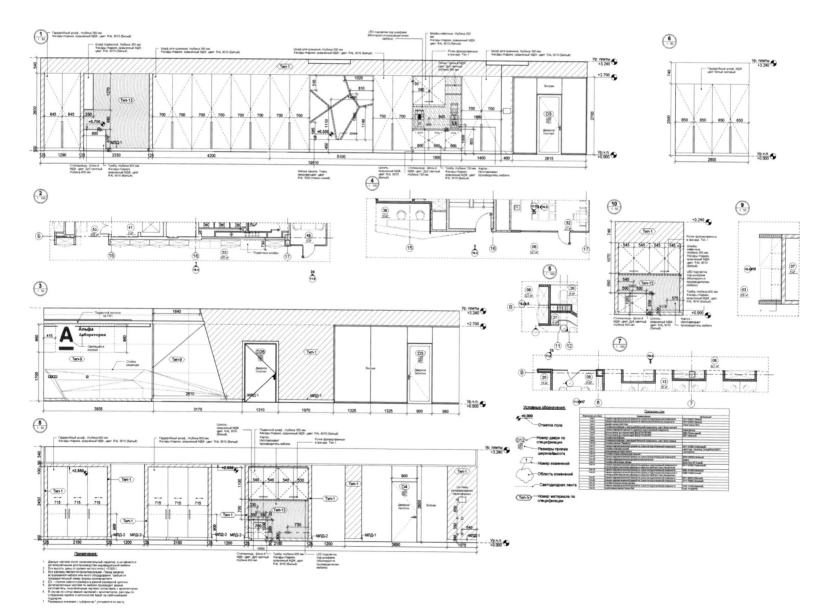

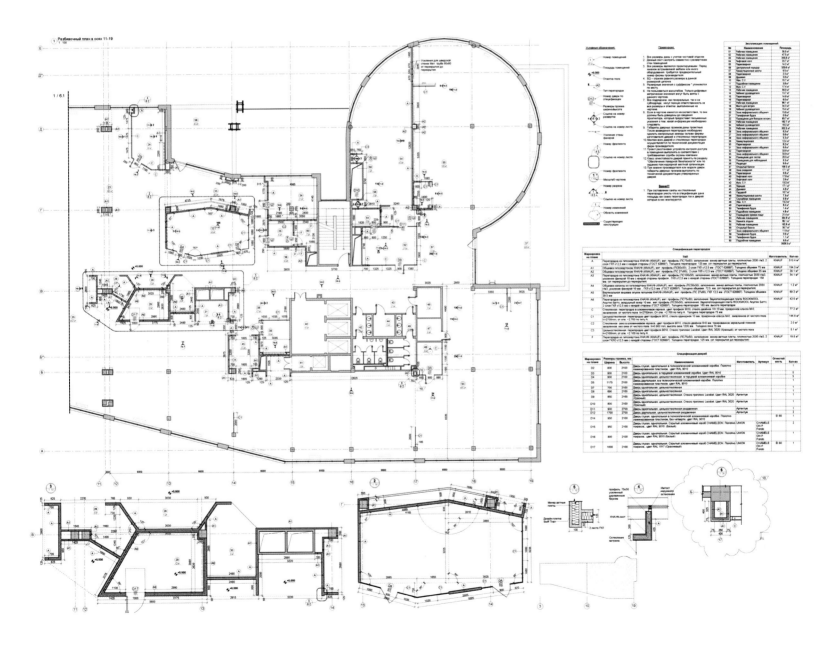

该办公室是由 IND Architects 工作室为阿尔法银行的一家支行设计的。首先，对于一个金融部门来讲，该创意是令人难以置信的。其次，它的功能性、醒目性和创新性使它成为阿尔法研究室员工的理想工作场所，这些员工年轻、有活力，专门从事阿尔法银行的电子商务业务。

室内的设计理念——超级英雄和街头艺术贴近员工生活。基于此，装饰材料选用了混凝土、木材、穿孔金属和各种各样的玻璃，如透明玻璃、磨砂玻璃以及图纹玻璃。亮丽的地毯引导着人们进入空间，会议空间中色彩缤纷，而公共空间中各式各样的圆圈能帮助员工找到相应的团体和部门。

办公室的墙壁具有以下功能：装饰性，夺目的漫画和超级英雄壁画覆盖墙壁，起到装饰作用；实用性，可用记号笔在墙上写字，部分墙壁上有公告板，供员工粘贴材料；信息公布和激励作用，如一些伟大人物的箴言。游戏区的天花板上装饰着人形图案，代表着团队精神，灯交叉设在人形图案中，意味着创意思维和新鲜想法。除此之外，这些手法也提高了空间的声音效果。

办公室的特色是员工不仅能在空间中工作，还能积极参加会议和"头脑风暴"，这也恰好定义了空间的布局：多会议空间和休闲谈论区（咖啡区）。游戏区中设有乒乓球桌，各种棋盘游戏以及玩飞镖的软垫墙。两条户外门廊中摆放了光亮的家具。开放式的会议大厅设在办公室中央，直接在墙上就能做展示。另外，各个会议室中都设有 LCD 屏。

多样化的灯光设计手法被应用到办公室中：公共空间的线形灯光、走廊中的LED背光灯以及游戏区和展示区中的底部照明灯。这是通过使用迷人的元素和有趣的细节来打造高档、动态和功能化的设计，使得空间不仅仅是办公室，更是有助于员工构思独特的现代策略的舒适空间。

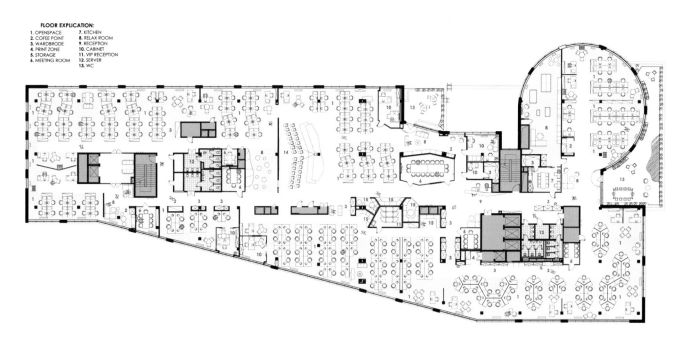

FINANCIAL/BUSINESS SERVICES

The Gores Group Headquarters

格雷斯集团总部

Architects: Belzberg Architects

Area: 11, 148 m²

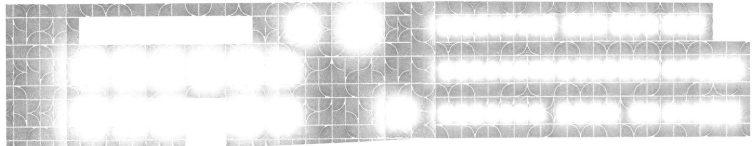

printed pattern on SGX interlayer
[follows slump pattern and dissipates around openings]

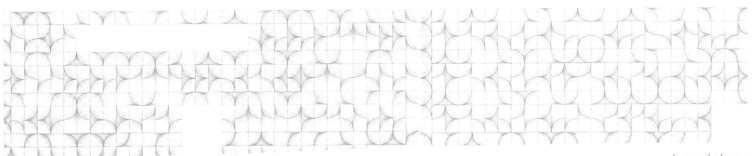

slumped glass panel
[pattern layout]

rendered elevation

glass composition ① 4mm annealed glass [slumped glass technologies]
② .06" printed SGX+ clear PVB
new stainless steel support ③
④ 4mm annealed glass [autoclave lamination]
existing building ⑤

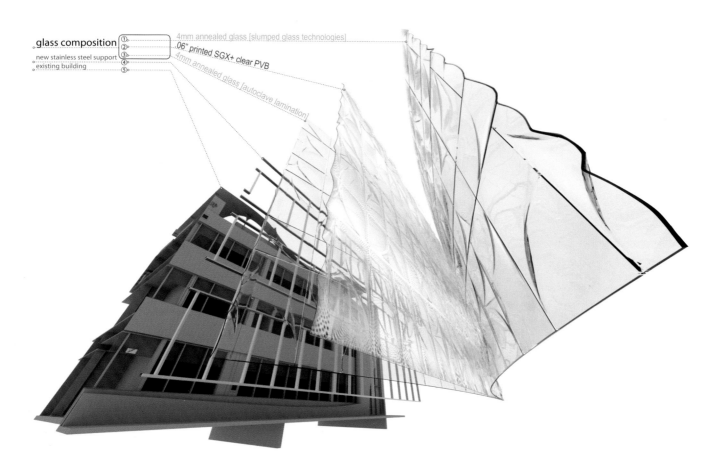

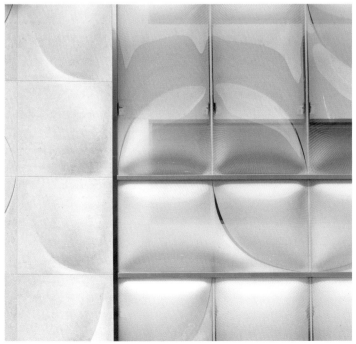

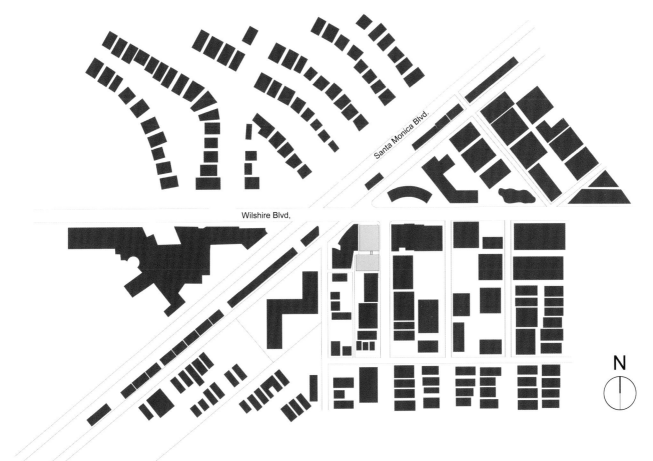

A sustainable strategy to reuse existing structures in lieu of demolishing and building new ones, greatly reduces waste and limits generating more (often unnecessary) building material. This is increasingly common practice with older commercial office buildings, often left desolate, which are in desperate need of upgrades. This ever growing challenge, inevitably presents the architect with the problem of making the older buildings economically viable to developers.

Our concept focuses on an alternative approach to renovating the standard commercial typology through the use of an innovative skin. A performative, custom made double-glazed facade system is utilized to manipulate visual opacity, acoustics, temperature, and program for a fundamental transformation.

The double layer facade system, made from custom slumped glass, helps to regulate indoor air quality in a seasonal fashion by mechanically venting hot air up and out in the summer and heating and circulating the cooler air in the winter. Carving out a new atrium space in the center of the existing structure mirrors this chimney effect from the facade by utilizing an operable skylight to release or retain heat depending on the conditions. Furthermore, the customized pattern interlayer sandwiched within the glass panels selectively filters views, privacy, and light based on specific site conditions and client needs.

该项目施行可持续性策略，重新规划了原有结构，而非拆除原建筑以建设新建筑。这便大大减少了浪费，消除了局限性，生成了更多（常被误认为是无用的）建筑材料。这种方式在老商业楼的改造中越来越普遍，被废弃的楼房往往急需升级。这种愈来愈难的挑战难免会是建筑师的难题：为开发商提供经济上可行的老商业楼改造。

设计理念专注于用替代性手法更新标准的商业楼，使之成为创新性商业楼。定制的明亮的双层玻璃墙具有隔音、隔热、不透、易改装的特点。

双层玻璃墙是由坍落玻璃制成，有助于随季节变化而调节室内空气质量，例如夏天热空气会上升并排出，而冬天则会使冷空气升温和循环。原结构中的新中庭空间具有烟囱般的效果，可开关的天窗有利于根据情况释放或维持空间的热量。根据场地条件和客户需求，夹在玻璃层之间的图案层选择性地阻挡了视线，起到了保护隐私和遮光的作用。

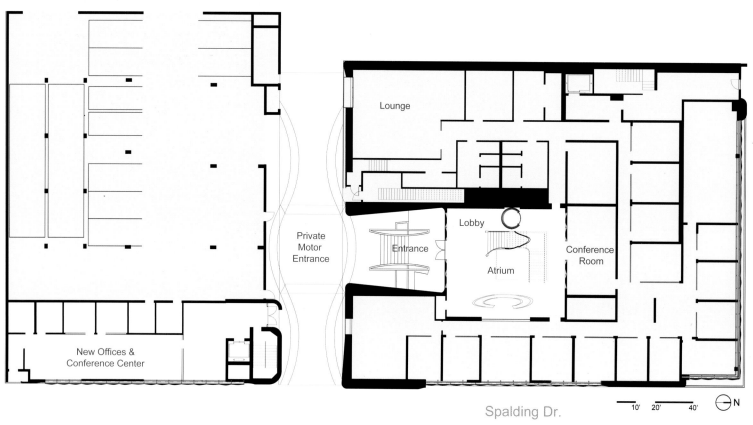

Co-Work Angel Office

Co-Work Angel 办公室

Interior Designer: PENSON

Client: CoWork

Furniture: Day 2 and A modern World

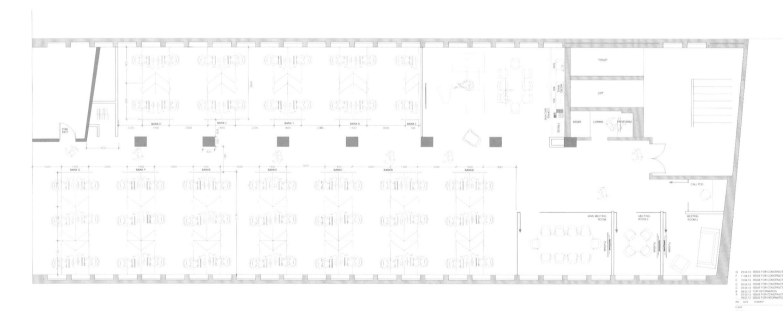

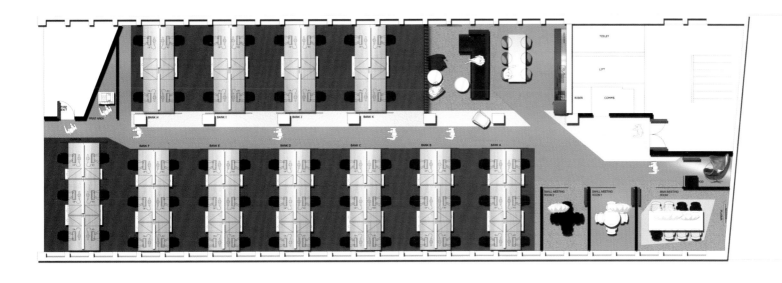

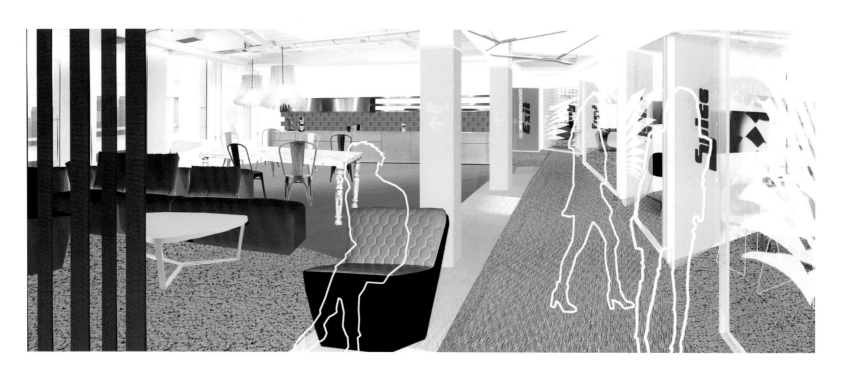

They've done it again. Leaders of architecture and interior design PENSON have created yet another office that people will actually look forward to going to work in.

Co-Work are providers of shared workspaces for entrepreneurs, start-ups and small businesses and have been taking the professional world by storm thanks to their unique, intelligent and economical business strategies and it's only a matter of time before they reach the top, with PENSON coming along to help make the journey as comfortable as possible.

No strangers to economical work processes themselves, PENSON have always been adamant that budget doesn't have to mean boring and regularly provide impeccably high standards of interior design and architecture to clients on a lean budget. Teaming up with a client with similar beliefs about always working to do the best for your customers was a no brainer and PENSON have now secured a rollout contract to design all of Co-Work's future buildings - starting with Co-Work Angel.

With an open brief in look and feel to design a sleek, modern space rejecting all things boring, they managed to create a spacious, vibrant and refreshing work environment whilst simultaneously ensuring nothing went to waste and that they kept to a shrewd budget.

After creating the large open plan office it was the finishing touches that would transform the room and add that extra bit of flair. Neon yellow paintwork, debossed ceilings, exposed pipework, impressive fluorescent lighting fixtures and glass conference rooms should have that covered. That is, glass conference rooms embellished with huge sets of angel wings - paying homage to the office's location.

Typically of PENSON, the workspaces are bright, spacious and aesthetically pleasing, the perfect example of their ability to make an area ooze character and style without being over exuberant. A separate rest area for workers to relax during their lunch break helps to bring a cozy contrast to the futuristic professionalism of the rest of the room, with a welcoming kitchen and top of the range coffee machine helping to keep everyone at the top of their game. This separate relaxing area is a real luxury for users of shared workspaces, with most other companies in the business providing one-room offices with no extra space to unwind.

Never the type to forget that happy employees are what make an office run smoothly, PENSON designed a special call pod to allow workers to make both personal and business calls in privacy and comfort rather than having to hang around in hallways or staircases like many other workspaces. They can even take a seat in the Bond villain-esque copper and leather armchair that was installed to add a little extra flair to the room.

The next step for Co-Work as they continue to make leaps and bounds in rapidly redesigning their field for the better is an expansion of their office in Borough that is soon to open its doors. After that it's on to the next plot as they carry out their impressive growth plans.

Overall this first project with Co-Work encompasses all that is PENSON: simple designs with a contemporary twist, not a boring light fixture in sight and clients taking all of their important calls from the comfort of a Bond chair.

PENSON 建筑和室内设计的负责人又成功创造了令人无比期待的办公空间。

Co-Work 专门为大企业、新公司和小企业提供工作空间，因其独特、经济、充满智慧的经营策略而享誉专业领域，并且，成为专业领域中的领头羊也只是时间的问题。然而 PENSON 的设计则让这个旅程更加舒适。

PENSON 熟知如何能让工作流程更加经济化，始终坚持低成本并不意味着枯燥的空间设计理念，而且他们经常以低成本为客户创造高标准的建筑和室内设计。本着相同的信念：始终毫不犹豫地为客户提供最好的服务，PENSON 与客户携手并进，且 PENSON 已与 Co-Work 签下合同，以 Co-Work 的 Angel 办公室设计为起点，负责该公司今后所有建筑的设计。

以设计视线开阔、开放式、井然有序的现代空间为宗旨，摒弃枯燥元素，创造一个宽敞、清新、充满活力的工作环境，同时确保在精细的预算基础上没有浪费。

大型开放式办公室的点睛之笔将转变空间，让空间更有格调。霓虹黄漆外观、凹式天花板、暴露的管道、荧光照明灯具以及玻璃会议室，一应俱全。玻璃会议室都用巨大的天使翅膀装饰着，符合房屋所处位置的特点。

对PENSON来讲，这个明亮、宽敞、美观的空间很有代表性，充分显示了PENSON的能力，让空间个性鲜明、风格独特、简而不繁。专门供员工午休的休息空间十分舒适，与其他未来派风格的专业空间形成对比；引人注目的厨房和高端的咖啡机让每个人都精神饱满。与业界许多只有单间办公室，没有额外的放松空间的公司相比，独立休息区对在公共工作空间内工作的人们来说，无疑是一种奢侈品。

PENSON深知，职员的满意度是公司能否顺利经营的关键，因此为职员设计了独特的通话间，令其不管是接听私人电话还是商务电话，都能保证隐私，舒适地通话，而无需跟其他办公室一样，在走廊或楼梯间接听电话。用铜和皮革制成的Bond villain扶手椅给空间增添了个性，职员可以坐在上面享受一番。

由于Co-Work以飞快的速度在进步，必须在领域内重新规划，做更好的自己。因此下一步就是在市镇开设新办公室，而且很快将投入使用。此后将在另一块场地上执行扩展计划。

总之，与Co-Work合作的首个项目完全体现了PENSON的风采：朴素的设计与现代的扭曲感；灯具醒目而不单调；客户还可以在Bond椅上接听重要电话。

FINANCIAL/BUSINESS SERVICES

Algomi

Algomi 办公空间

Design Agency: Resonate interiors

Furniture: Design Brokers and Senator international

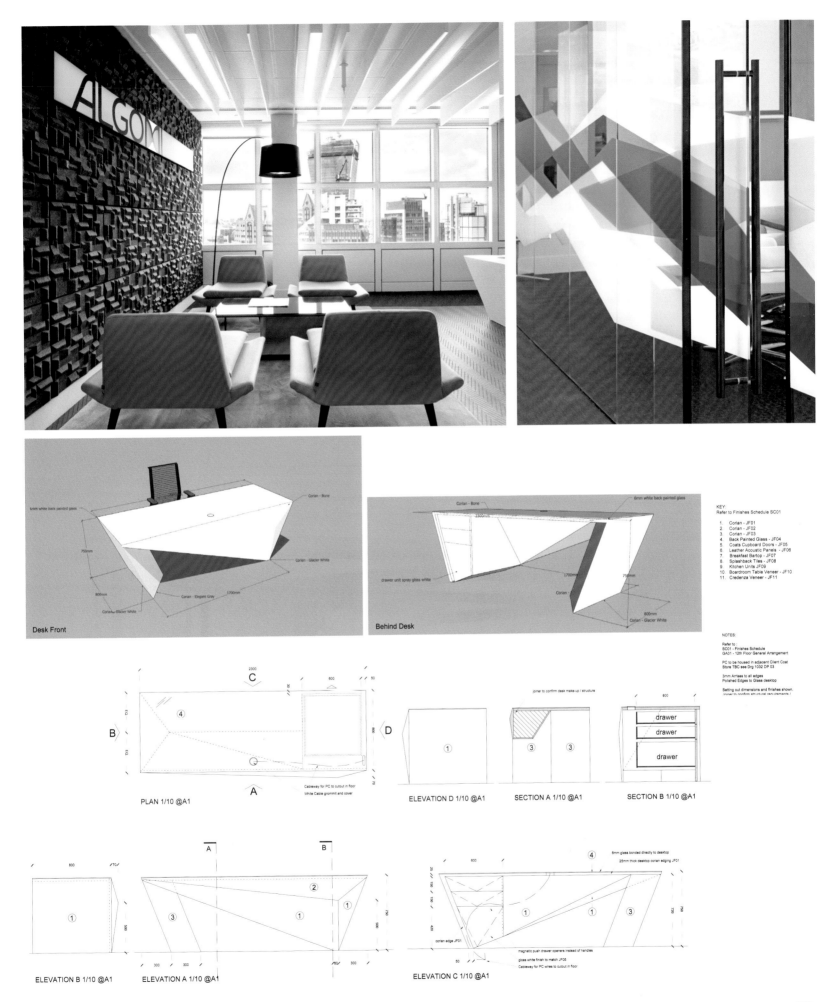

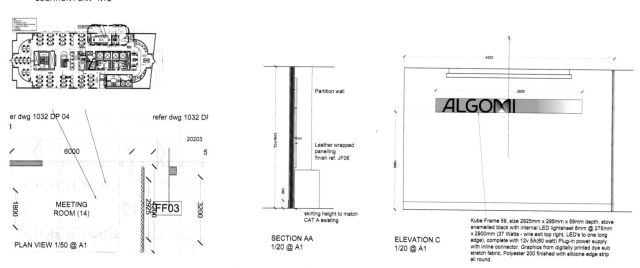

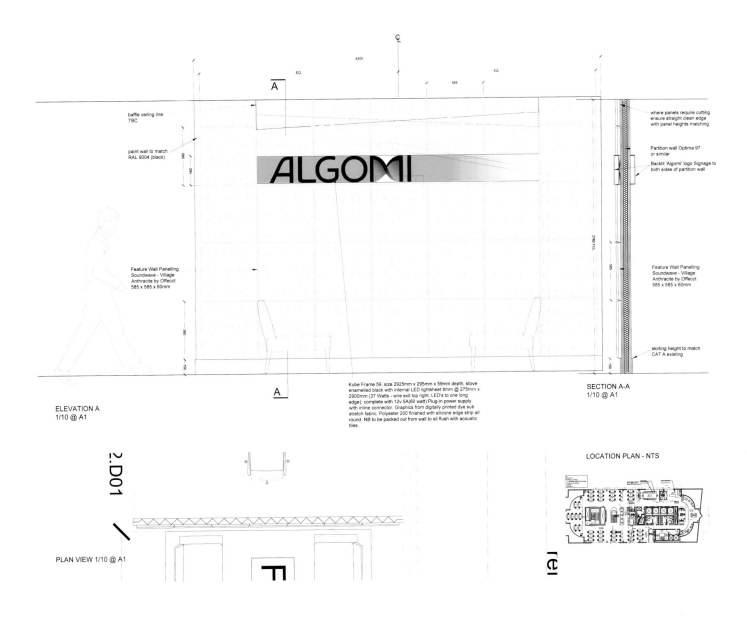

Resonate Interiors design and delivers new contemporary offices for Algomi.

Resonate Interiors were approached by Algomi to design their new London HQ. Algomi have enjoyed exponential growth over the last 12 months, having found new premises at 1 America Square in the City of London.

Algomi delivers a laser-like focus on trading velocity in Fixed Income. Clients include Global Investment Banks, Exchange Providers, Brokers, Platform Providers and Investor Client Firms.

The brief was to house 80 trading style desks within a contemporary environment that reflected the ethos of this energetic and visionary business. The design concept revolved around the twisting Algomi logo design. The scheme is essentially black and white with a hint of the Algomi turquoise blue. The juxtaposition of hard and soft materials was extremely important due to the limited palette.

Michael Schmidt, Chairman of Algomi commented: "The new office expresses exactly the vision and values of Algomi in creating an inspiring and open minded atmosphere, were employees feel home and taken care of. Taking the Algomi brand into consideration and translate it into our new office has clearly been achieved by the excellent designers work and quality execution."

On entering the 12th floor, visitors are greeted with a beautiful handcrafted Corian reception desk surrounded by Offect acoustic panels and a bespoke flowing ceiling. The reception adjoins a well-appointed business lounge with a plenty of work settings, to allow visitors work and relaxation space.

The Boardroom has a stunning fourteen person board table handcrafted in smoked oak with an inset glass panel and credenza to match. The workplace is open plan, enjoying the stunning views of the city, with black and white desks by Senator. In the heart of the floor is a break out space with a 14 person stone breakfast bar, supported by low level seating area and 3 large plasma screens reflecting the business that Algomi are in. The bar area itself, is designed with great detail in mind, using Domus matt black tiles creating artwork in themselves.

Timescales were fast and furious! With only 2 weeks design and 4 weeks on site, working alongside the contractor's Kingly, the project was executed to a high standard. Every detail thought through from the bespoke artwork panels to the impressive elongated breakfast bar.

In conjunction with Design Brokers from Denmark, furniture was specified again to reflect modernity, clean lines and classic design. Great attention to detail was applied by Resonate in specifying a range of fabrics and finishes that allowed a sense of luxury and style and brought the interior scheme to life.

"Every detail on this project was specified down to the cups and saucers", Pernille Stafford of Resonate Interiors added, "It was such a joy to work with a fantastic energetic client that made quick and great decisions along the way. Overall a fascinating & very rewarding project and one in which Algomi continues to thrive, with another floor already being planned."

Resonate Interiors 为 Algomi 设计了现代、全新的办公空间。

Resonate Interiors 应 Algomi 邀请，为其设计新伦敦总部。Algomi 在上个年度迅速发展，在伦敦金融城美洲广场定下了办公室新址。

Algomi 高度专注于固定收益中提升交易速度。其客户包括全球投资银行、兑换商、经济商、平台提供商和投资者客户公司。

本案的目标是将 80 款交易桌放置在现代的空间环境中，这种现代环境恰好与充满活力和远见的业务相一致。设计理念围绕着弯曲的 Algomi 商标设计。用色方案主要采用了黑色和白色，并用 Algomi 翠蓝色稍加点缀。软硬材料的并用在有限的色彩方案中发挥了重要作用。

Algomi 董事长麦克·舒米特评论道："新办公室将 Algomi 的憧憬和价值表现得淋漓尽致，具有激发灵感、开放思想的氛围，赋予了职员家一般的感觉，也让他们感到颇受关注。将 Algomi 的品牌融入设计，用新的办公空间传达其意，这一目标的实现得益于设计师出色的工作和质量监控。"

进入第 12 层，美观的手工可丽耐接待台便呈现在访客眼前，被隔音背板和定制的流动天花板环绕起来。旁边便是配置齐全的商务休息室，大量的工作设备能满足访客工作的需求，当然也可用作休闲空间。

董事会会议室中摆放着设有 14 个席位的高级会议桌，由烟熏橡木手工制成，嵌有玻璃板，并配有橱柜；在开放式工作区中能欣赏迷人的城市风光，其间摆放了由 Senator 制作的黑白工作桌。楼层中央是休息空间，设有能容纳 14 人的石餐台，较之更低的是座位区，3 个大型等离子屏幕展示着 Algomi 正在参与的业务。吧台区设计十分精细，Domus 暗光黑色瓷砖的铺砌，本身就是一个艺术品。

完成项目的时间比较紧迫，只有 2 周的设计时间，另有 4 周的时间来参与现场工作并与承包商协作。项目遵循了高标准，从定制艺术壁板到长餐台，每个设计都细致入微。

办公室的家私采用了现代风格，通过与丹麦 Design Brokers 的合作，融合了清晰的线条与经典款式。Resonate 注重细节，空间的外观和采用的布料无不透露出一种豪华感，且风格独特，赋予了室内空间活力。

"项目的每处细节，包括杯盘茶碟，都十分细致"，Resonate Interiors 的佩尼莱·斯坦福补充道，"能与这样充满活力的客户合作，是件乐事，且项目进展快，决策也较迅速果断。这个项目值得付出，打造出的效果可谓引人入胜。Algomi 仍在继续努力，已有另一楼层在规划中。"

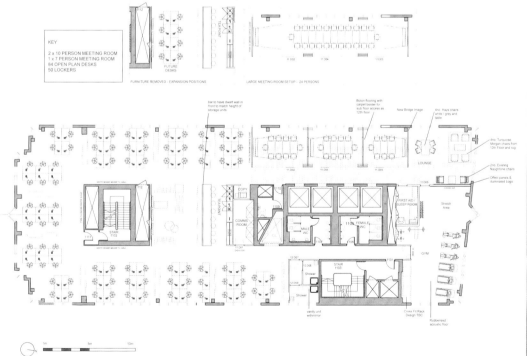

KEY
2 x 10 PERSON MEETING ROOM
1 x 7 PERSON MEETING ROOM
84 OPEN PLAN DESKS
50 LOCKERS

FINANCIAL/BUSINESS SERVICES

ZS Associates

致盛咨询公司办公室

Design Agency: DSP Design Associates Pvt. Ltd.

Location: Gurgaon, India

Area: 12,653 m²

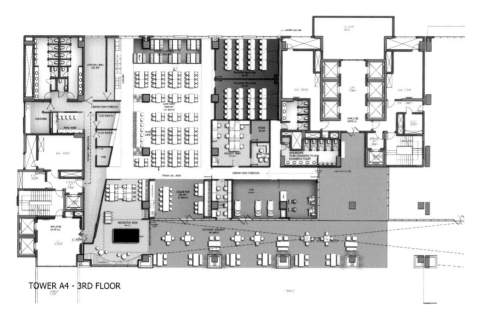

TOWER A4 - 3RD FLOOR

TOWER A4 - 4TH FLOOR

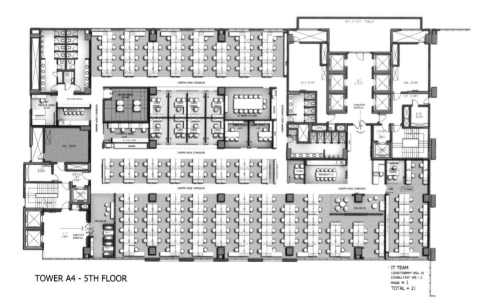

TOWER A4 - 5TH FLOOR

ZS is a professional services firm that has broad and deep expertise in the 2 areas that create customer demand: sales and marketing. Their services include go-to-market strategies based on customer insight and product positioning analytics, as well as business strategy implementation based on tactical planning and deployment of operational effectiveness levers such as sales compensation, customer management strategies and tools, customer data management and analytics, and decision support platforms, to name a few areas. ZS consultants work closely with the clients across these areas to produce maximum returns, creating impact where it matters.

To achieve the best possible workspace in ZS's Gurgaon Capability and Expertise Centre, DSP categorized their work modes into 4 types:

· Focus: ensures concentrated and uninterrupted solo effort.

· Collaborate: ensures brainstorming with others on board to foster innovation.

· Learn: ensures absorption of new skills, augmentation of knowledge through formal or informal training.

· Socialize: ensures interaction in the right manner in the right space to build value through open communication.

This array of formal and informal spaces is designed to facilitate the work modes:

Focus	Collaborate	Learn	Socialize
Work Desks	Collaboration Hubs	Training Rooms	Breakout Areas
Cabins	Meeting Rooms	Meeting Rooms	Cafeteria
Focus Rooms	Booths		Lounge
Booths	Focus Rooms		Recreation Area
	Lounge		Gym

The entire office is designed to be flexible and scalable, by using the workstation as a building module. Some important philosophies guided the design:

· Creating a modern, yet warm and inviting feel. ZS Delhi office incorporates natural materials including stone, wood and plants. Shared spaces feature different looks and layouts from restaurant-style booths to mid-century modern furniture. Bright colors infuse the spaces with energy. The workstations are aligned toward the windows to get access to maximum sunlight.

· Fostering collaboration. With an "inside-out" floor plan, workstations are at the perimeter of the floor and cabins—or private offices — and meeting rooms are on the interior. Partitions between individual work stations are low and can also be removed or shifted to create flexible, collaborative workspaces. The workstation can be rearranged to fit a new work setting while ample open space comes with scope for expansion in future, adding more strategic viability to the space. There are numerous large glass whiteboards to foster brainstorming and innovation.

The cabins and meeting rooms are situated in a way so as to ensure easy accessibility from all areas of the floor. Each floor is well-equipped with comfortable informal spaces such as breakout zones, collaboration hubs, booths and focus rooms.

· Emphasizing what defines ZS' business, culture and people. The

space features photographs of real ZSers to add to the authenticity—we are what we express—and motivational quotes line the walls. A gym, recreation room and cafeteria promote wellness and socialization. Linear benching is used to make the workspace more efficient and agile. An inviting reception lounge welcomes visitors. In a nod to ZS' global clientele and working model, meeting rooms and collaboration areas are named after regions around the world.

ZS' wall of inspiration is the best manifestation of the collaborative effort between all the teams. This striking space features messages from various ZSers from around the globe about their work, experiences and aspirations. The entire office sports motivational quotes from famous leaders.

ZS's team was very meticulous about the development of space concept and the visual appeal of the office. The Office Managing Principal, Stephen Redden, was extensively involved with the DSP team to ensure that the new ZS office echoes the ZS culture, character and philosophy. ZS's global branding partner, "Branding Business", developed a series of graphics for all the spaces around the office, to ensure that the internal branding of the Gurgaon office was in line with ZS's Global offices.

Branding Business was also instrumental in closing on the color palette to establish ZS' brand identity in their workspace. DSP, in tandem with ZS, decided upon a tranquil primary palette of black, white, gray—supported by wooden—and secondary vibrant color palette of orange, rust, teal and purple, which align with the firm's chosen brand and presentation colors.

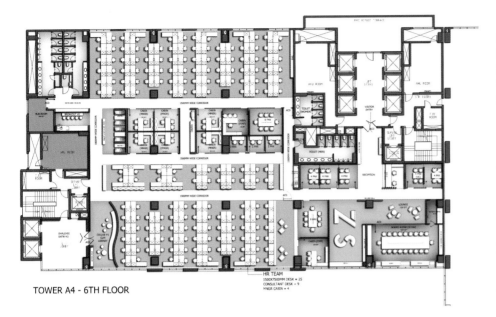

TOWER A4 - 6TH FLOOR

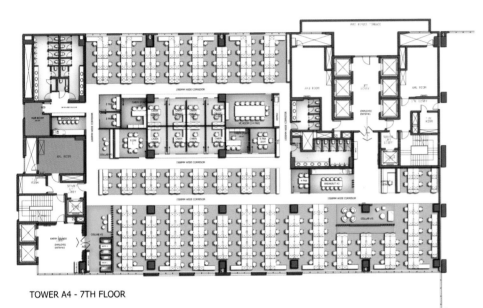

TOWER A4 - 7TH FLOOR

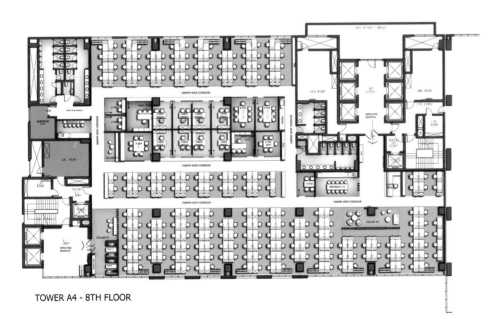

TOWER A4 - 8TH FLOOR

致盛咨询公司是一家专业的服务公司，专注于为销售和营销领域的客户提供多项专门服务，其中包括依据客户想法和产品定位分析制定走向市场的策略，以及基于战略性计划和运营效益杠杆制定商业策略实施计划，如销售报酬、客户管理策略和方法、客户资料管理和分析以及决策支持平台等。致盛咨询公司在这些领域与客户深入合作，以帮其实现收益最大化，同时在这些领域产生了重要影响。

为了给致盛古尔冈人才中心打造最好的办公空间，DSP 将这些人才的工作模式分为 4 类：

· 专注：注意力集中的、不被打断的个人工作；

· 协作：与他人一起"头脑风暴"，鼓励创新；

· 学习：学习新技能，从经验、交流和正式或非正式的培训中获取工作知识；

· 社交：与人交流，建立有利的人际关系，组建团队，平衡工作与生活。

针对上述工作类别设计的正式空间和非正式空间：

专注	协作	学习	社交
工作桌位	合作中心	培训室	茶歇空间
封闭小间	会议室	会议室	自助餐厅
单人工作间	电话间		休息厅
电话间	单人工作间		娱乐室
	休息室		健身房

办公室设计采用了工作站来组建建筑模块，具有灵活性和可扩展性。以下是指导设计的核心理念：

· 营造现代感、温馨感和宜人感。这个位于德里的致盛办公室采用天然材料装饰而成，如石材、木材和植物。共用空间的布局和外观各不一样，如餐厅风格的座位和中世纪现代私家具。空间色彩缤纷，充满活力。工作座位都沿窗户摆放，光线充足。

· 促进协作。由内至外的布局巧妙地把工作台设在了楼层的边缘，而封闭小间（个人办公空间）和会议间则设在内部。单人工作台位之间的隔板较低，也能通过拆除或移动来形成更为灵活和有利协作的工作空间。工作台也可重新布局成新工作环境。此外还留有充足的空间以便未来发展，这也增加了空间的战略发展可行性。另外，许多的玻璃白板则是为进行"头脑风暴"和创新活动准备的。

封闭小间和会议室布局巧妙，不管从楼层的哪个方向都能进入到这些空间。每个楼层的设备都十分齐全，还设有休闲空间，如休息区、合作间、电话间和单人工作间。

· 突出致盛的服务、企业文化和人文关怀。空间内用致盛成员的照片和励志箴言装饰着，增加了空间的真实感：展示的风采就是我们的风采。健身房、娱乐室和餐厅鼓励健康生活和社交活动。长凳的设置增加了空间的使用效率和灵活性。迷人的接待室欢迎着访客的到来。会议室和协作区以世界各处的地名命名，体现了致盛的业务和客户的全球性。

致盛灵感墙恰到好处地体现了团队之间的友好协作关系。这个令人惊讶的空间展示了来自世界各地的致盛成员对工作、经历和灵感的感言。而整个办公空间则展示了著名领导人的励志名言。

致盛团队十分注重办公空间的设计理念和视觉效果。办公室管理负责人 Stephen Redden 特意参与到 DPS 团队中，确保致盛办公室与致盛的文化、特征和理念的一致性。致盛在全球的品牌合作伙伴"Branding Business"为办公室的所有空间设计了一系列的图片，以确保古尔冈办公室内的商标与其他的致盛全球办事处一致。

"Branding Business"还在色彩选择上提供了帮助,更好地让空间体现致盛的品牌身份。DSP 和致盛最终决定采用黑、白、灰三种为主色,辅以木色和鲜艳的橙色、锈色、青色和紫色。这些都是能体现公司品牌和特征的色彩。

FOOD/BEVERAGE

Red Bull - Cape Town Headquarters

红牛开普敦总部

Design Agency: Giant Leap

Area: 1,600 m²

Photography: Adam Letch

Red Bull, the highest selling energy drink in the world put out their wings to move to new premises at the V&A Waterfront.

The importance was placed on understanding their brand and culture. "We are looking for a creative, yet functional space where our staff can work individually as well as collaborate as a team".

Giant Leap Workspace specialists were brought in to understand what this meant and what they wanted to achieve out of their 2-storey office space. We studied their staff and how they worked. How much time they spent in the office and what the tasks at hand were when in the office.

The 2-storey, 1,600 m² space was designed around their people and their brand. We introduced walls showcasing their incredible action packed adventurers, racing games for their staff and war rooms for them to strategise in. The brief was high end, yet funky that embraced them as a global brand. It also needed to fit into their budget. The space was designed around 74 staff allowing for growth of up to 15 people.

Technology was an integral part of the design and brief. They communicate globally as well as with their Johannesburg office. We designed the space around an industrial look with no ceilings and all the services exposed, taking into account that the offices need to be acoustically sound proofed. The floor space was populated between 2 floors in a way that encourages their staff to move between the floors and not creating any hierarchical structures.

The kitchen area was designed for staff lunches and collaboration but doubles up into an area that can be transformed for entertaining and a DJ, in true Red Bull spirit.

It was important to work with the natural light and let it flow through the office environment.

The final product is a space that allows for collaboration and privacy to co-exist in a noise controlled environment. Cutting-edge design displays their brand throughout.

The space showcases the sporting events and talented athletes that Red Bull sponsor as well as displaying their product.

Interactive spaces where staff can integrate, socialise, collaborate and compete.

We created a world class "war room" for people to strategise and think out the box.

The new office integrates technology with a variety of different zones allowing their staff to be mobile.

全球能量饮料销量最高的红牛公司搬迁至阿尔佛莱德码头的新办公室中。

设计的重点是理解红牛的品牌和文化。"我们在寻找一个既有创意,又有功能性的空间,职员能够在这个空间里独立工作,又能与团队合作。"

Giant Leap 设计机构的办公空间专家接受了邀请,并深入思考了这些要求,以及如何设计这两层办公空间的问题。我们观察了公司的职员及其工作方式,例如他们在办公室工作的时间;当他们在办公室时,处理的是什么类型的工作。

这两层空间的总面积为 1 600 m^2,空间是围绕着职员和公司品牌而设计的。例如,墙壁上展示了参与刺激的赛车冒险的职员,赛车游戏可供职员休闲,策划室则是为制定策略准备的。设计方案必须与红牛的全球品牌战略相符,空间必须高端时尚,花费也要限制在预算范围之内。空间是按照 74 人所需的空间设计的,此外还有多于 15 人的备用空间。

科技是设计中不可分割的一部分。技术允许该办公处和全球的办公室相联系,也包括位于约翰内斯堡的办事处。空间的设计采用了工业风格,未设天花板。另考虑到空间的隔音效果,所有的服务设施也都裸露了出来。两层楼间的空间设计旨在让职员能够在楼层间自由活动,而不产生任何等级感。

厨房空间是由职员就餐空间和合作空间组成的,能够转变成娱乐空间和 DJ 房,这与红牛的精神十分契合。

在充满自然光线的空间中工作也同样重要。

本案最终便形成了一个允许团队协作和独立工作共存的空间,而且能较好地控制噪音的工作环境。这种高档的设计充分展现了公司的品牌形象。

空间展示了红牛赞助的运动赛事和杰出运动员,还展示了企业的产品。

在互动空间中,员工可以相互交流、合作与竞争。

我们创造了一个顶级的"策划室",有助于职员制定策略和创新性思维。

新办公空间采用了集成科技,令职员可以在各种空间内活动。

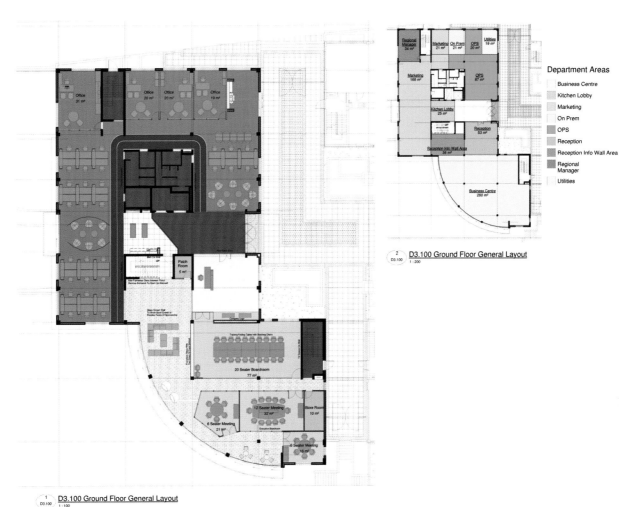

FOOD/BEVERAGE

Head Offices, Red Bull Canada

加拿大红牛总部

Design Agency: Johnson Chou Inc

Client: Red Bull Canada

Floor Area: Music Academy & Headquarters: 880 m² total

Extension Area: 465 m²

Location: Toronto, Canada

Photography: Tom Arban Photography

Located in downtown Toronto's Queen West Art District, the design concept for Red Bull's Headquarters is derived from the notion of spaces as vessels for transformation - for both personal transformation and as receptacles that capture and record one's personal development.

The Music Academy is one vessel that encapsulates Red Bull's corporate mantra of "mentorship instead of sponsorship", functioning as a forum for the exchange of ideas - an inspirational hub where knowledge and experience are gained and disseminated. Recording these moments in time, the design features a "Memento Wall" reaching 2-storeys tall in the atrium. The wall is affixed with clear, acrylic boxes containing objects, writings, and other ephemera that were inspirational to participating artists during their time at the Academy.

The Red Bull space was later expanded by 465 m² and focuses on 4 main components: a bar/lounge area, a linear open workstation concept, a wool felt-clad form, and a boardroom. The client's brief was essentially three-fold: to create an inspiring space for staff in administrative/accounts who are normally relegated to bland, impersonal spaces; to predominantly utilize materials from a reclaimed source; and to create a space consistent in concept and form to the original, yet be a reinterpretation of it.

该项目位于多伦多皇后西街艺术区市中心，其总部设计理念源自一种把空间比作"容器"的概念，它能适应转化和自我转化，还能捕捉和记录个人的发展过程。

音乐学院便是一个能够压缩装下红牛的企业箴言的容器：与其做一名赞助者，还不如当别人的良师益友。音乐学院相当于一个论坛，人们可以在其中交流想法，它就像是灵感的源泉，知识和经验都能在这里获取与分享。为了记录一些特殊的时刻，该项目在中庭设计了一道两倍高的"纪念墙"。墙上粘着透明有机玻璃盒子，盒子里面装有物品、字条以及一些闪现在脑海的东西，这些东西让当时参与学院项目设计的艺术家充满灵感。

红牛办公空间后来被扩大了约465 m²，主要集中在4个方面：酒吧（休息）空间、会议空间、线形开放式办公空间以及给人以舒适感的装饰。客户对空间的要求分为3层：为通常被分配到平淡无奇的工作空间的行政人员和财务人员创造一个鼓舞人心的空间；充分利用回收材料；创造一个在理念和形式上与原空间保持一致，却又重新阐释了空间的设计。

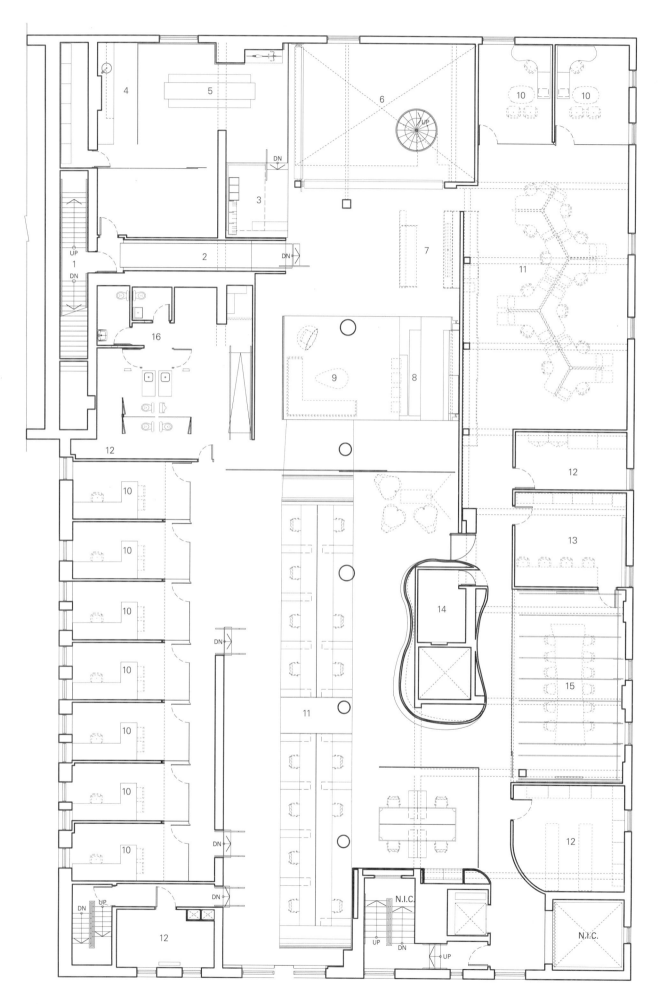

1. ENTRANCE STAIR
2. TRANSITIONS TUBE
3. WAITING / LIBRARY
4. KITCHEN
5. DINING / LUNCH AREA
6. ATRIUM
7. RECEPTION DESK
8. BAR MEETING AREA
9. LOUNGE MEETING AREA
10. PRIVATE OFFICE
11. OPEN WORKSTATIONS
12. PRINTING / STORAGE
13. TEAM OFFICE
14. SERVER ROOM
15. BOARDROOM
16. WASHROOMS

FOOD/BEVERAGE

Coca-Cola London Headquarters

可口可乐伦敦总部

Design Agency: MoreySmith

Client: Coca-Cola

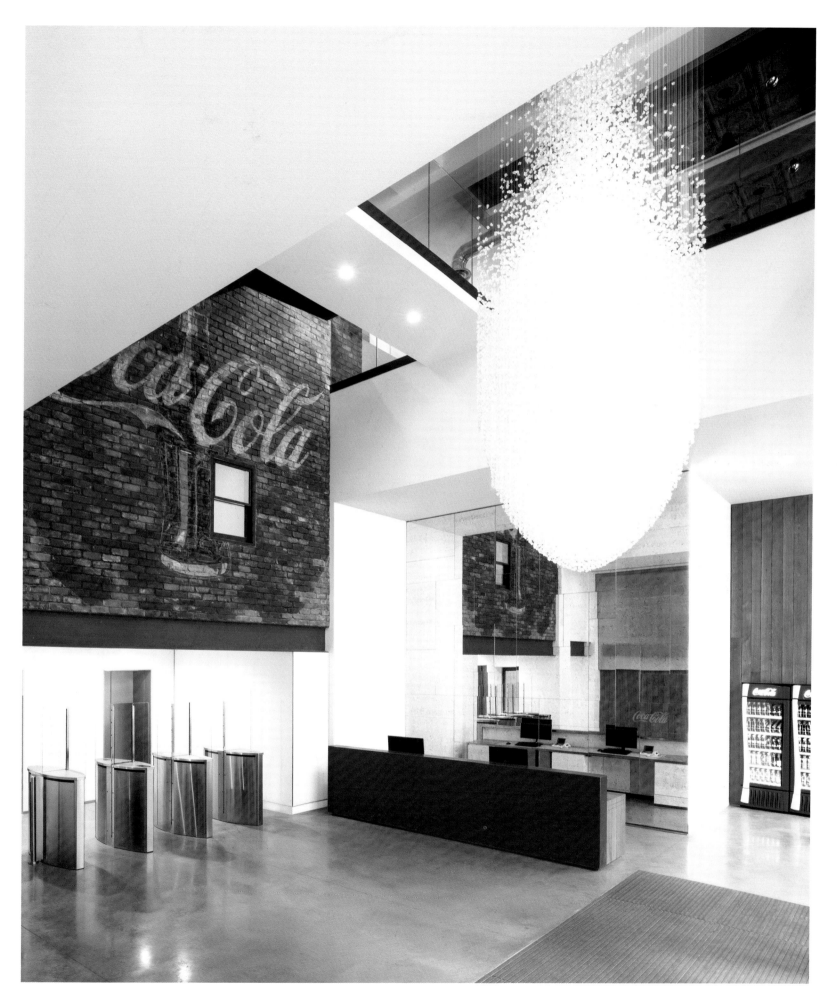

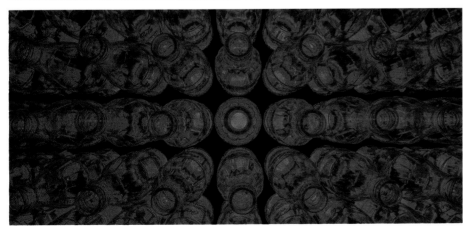

Coca-Cola's UK subsidiary has moved into its new London headquarters in Wimpole Street, London W1, Relocating 300 Coca-Cola employees from Hammersmith, W6.

The company's new London base is a 1920s building which has been refitted throughout. It is set over four floors, covering 6,132m² and features a roof terrace, café, meeting rooms, open plan office space and hot desk facilities.

The interior of the building is a celebration of Coca-Cola's heritage and place in popular culture. The design unites the Edwardian Baroque-style front of the building, with a more modern rear extension via a staircase which runs through the core of the building. A custom-built, double-sided display wall has been installed over 3 floors and provides a spectacular backdrop to display the brand's long heritage, including iconic original Coca-Cola memorabilia from its archives in the US.

可口可乐英国子公司搬至位于伦敦西一区温坡街的新总部，重新安置了原来在西六区哈默史密斯工作的 300 名职员。

新总部是由一栋 20 世纪 20 年代的建筑全面改造而成。建筑分为 4 层，占地面积约为 6 132m²，设有顶楼露台、咖啡区、会议室、开放式办公室以及公用办公桌等设施。

建筑中的室内设计是对可口可乐传统和流行文化的一种赞美。建筑外观是爱德华时代巴洛克建筑风格，而建筑里面从楼梯到建筑中心的延伸部分采用了现代风格。定制的双面陈列墙在三层楼上都可看到，成为了品牌展示的背景墙，其中包括从美国带来的标志性的可口可乐纪念品。

FURNITURE / PRODUCT

One Workplace Headquarters

One Workplace 总部

Architects: Design Blitz

Location: California, US

Area: 3,251 m²

Photography: Bruce Damonte

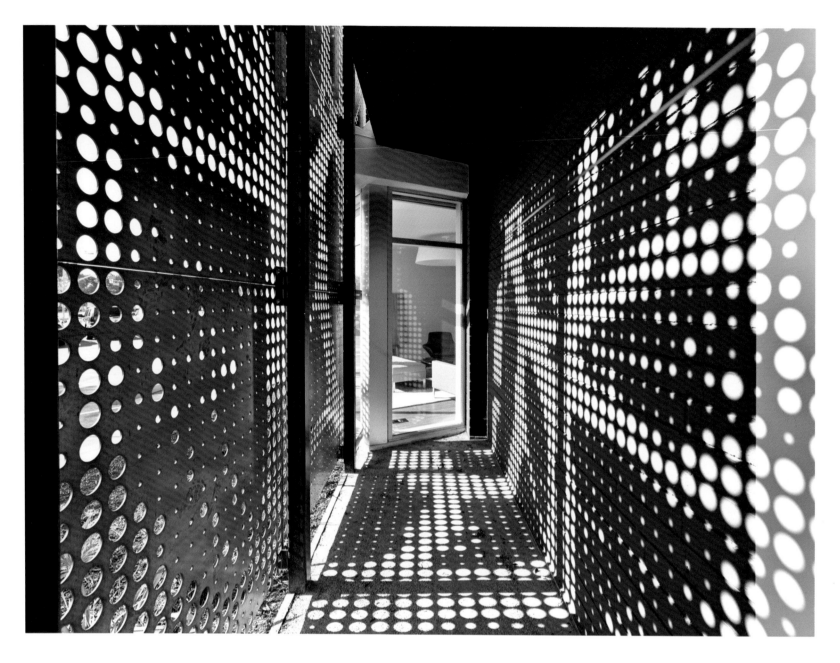

The One Workplace Headquarters project was a unique opportunity for both One Workplace and Design Blitz. One Workplace is the largest furniture dealer in Northern California and the single source for Steelcase furniture in the San Francisco Bay Area. The ambitious directive was to re-define the architectural standard not only of the company, but also of the showroom experience itself and create a bleeding edge, world-class workplace that serves both employees and customers alike. One Workplace had already shifted the industry paradigm of sales and showrooms away from a transactional experience to one of collaboration and partnership. This leadership position makes One Workplace a formidable client partner, and the next iteration of their corporate headquarters needed to embody this innovation and progression. No longer a static showroom, the working showroom needed to demonstrate what is possible when great minds come together within the context of a multi-disciplinary design lab.

Design Blitz was a natural design partner for One Workplace. As a young firm who is redefining the architectural service model through their use of client-friendly technology and open platform, they had already established a track record for creating innovative workplaces for several of Silicon Valley's most successful technology companies including Skype, Comcast and Square. Working in partnership with One Workplace, Design Blitz realized the transformation in built form, creating a working showroom representative of One Workplace's evolution.

The project consists of 3,251m^2 of office/showroom/workspace with an adjacent 16,723m^2 warehouse (warehouse improvements were completed separately). In addition to the warehouse, the site also included an existing 929 m^2 stand-alone, mid-century office building layered with many years of dated tenant improvements. The project successfully connected this building and 2,323m^2 of warehouse into the re-imagined workplace. Design Blitz's design for the new facade and landscape improvements expand One Workplace's space into a multi-functional indoor-outdoor environment.

In addition to redefining the showroom experience, the project is also about adaptive reuse and urban regeneration. The site is located adjacent to the San Jose Airport, arterial roads and a freight train station in an industrial part of Santa Clara. The concept of transit innovation with the omnipresent planes, trains and automobiles provided significant inspiration for the design team. The warehouse and office buildings were both in disrepair. After exploratory demolition, Design Blitz determined that the small office building had great bones that could be salvaged and celebrated in the new project. Given the raw nature of the existing warehouse and office building and the exploratory nature of the Design Lab work typology, it was determined that the design vernacular should avoid the pristine

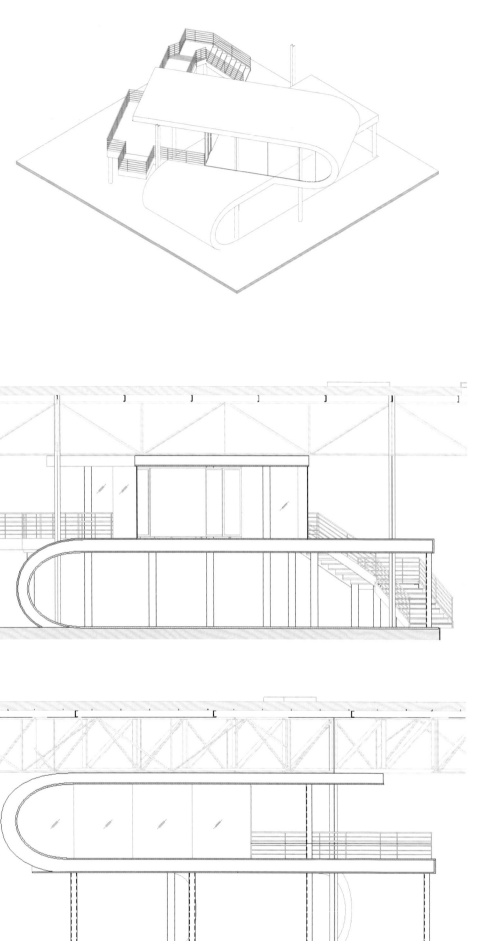

and celebrate the industrial history of the site.

During initial conversations, concepts of the kitchen as the family hub and the dining table as key elements kept emerging. As such, Design Blitz used the kitchen as an opportunity to bring all users together and as the programmatic connection between the 2 buildings. Quite literally, the office meets in the middle. Upon entering the building you are immediately presented with the work-cafe. It is an area to meet and eat. Leading with this hospitality function ensures that customers and users encounter a warm and welcoming space. It was also during these initial conversations that Design Blitz began mapping both the client and user experiences through the space. The elevated conference room and observation platform allows members of the One Workplace team to quickly survey the floor and show customers how a variety of systems solutions can intermix to create a unified, flexible and layered approach to workplace layout.

One Workplace occupied their previous office for over a decade and the new headquarters needed to last at least that long. Design Blitz planned for longevity and flexibility by providing a raised floor system in the open office for easy future furniture reconfiguration as well as limiting color and pattern to elements that are easily interchanged as trends change.

In learning more about how One Workplace engages with their clients Design Blitz determined that environmental context is key to effective sales. Customers will purchase an entire space, an environment rather than a single piece. It was crucial that the architecture supported the furniture and not the other way around. Individual spaces were designed holistically to encourage an emotional connection by the customer.

In addition to being an innovative design, the project demonstrates strong metrics for the economics of efficiency. One Workplace moved from a 4,181m^2 space into the new 3,252m^2 space while increasing staff from 101 to 165. The increased efficiency was achieved by reduction in workstation foot print and a move by the majority of the sales team to a mobile work flow where workers do not have a dedicated workstation. Mobile workers store their belongings at a centralized location and work either at a shared workstation or in the soft seating of the work cafe or alternate work areas. One Workplace is walking the walk when it comes to modern work typologies.

该总部设计项目对家具经销商 One Workplace 和 Blitz 设计来说都是一次特殊的机会。One Workplace 是北加利福利亚最大的家具经销商，也是旧金山海湾地区唯一一家 Steelcase 家具的供应商。项目的目标设定显示了它们的雄心：重新定义建筑标准不仅仅是为了公司，也是为了展厅自身给人带来的体验，更是为了给职员和客户创造一个顶级的工作场所。该公司已经完成了从以展示和销售为交易程序的产业模式向建立合作伙伴关系为业务模式的转型。One Workplace 的领导地位使它变成一个强大的客户和合作伙伴，而且，接下来的公司总部改造也需要体现公司的创意和进步，而不是仅仅停留在静态的展厅上。展厅必须体现一点：当大胆的思维相聚在多学科设计实验室中时，一切皆有可能。

Blitz 设计自然是 One Workplace 最好的设计合作伙伴。作为一家年轻的公司，它用有益于客户的技术和开放的平台重新定义了建筑服务模式。Blitz 设计已经为几家位于硅谷的成功的科技公司设计了创意办公室，其中包括 Skype、Comcast 和 Square。在与 One Workplace 的合作中，Blitz 设计完成了建筑形式的转型，创造了一个体现 One Workplace 的发展历程的工作空间。

该项目是由约 3 251m^2 的办公室、陈列室和工作区，以及相邻的面积约为 16 723m^2 的仓库组成（仓库升级单独完成）。除仓库外，场地上还有一栋占地约 929m^2 的独立建筑，这栋建筑建于 19 世纪，还留下了多年前租户所做的翻修痕迹。该项目成功地将这栋建筑和约为 2 323m^2 的仓库连接起来，构成办公区。Blitz 设计让办公室焕然一新，并对周围景观做了改进，将办公空间扩建成为一个多功能的室内外环境连为一体的空间。

因为场地靠近圣荷西机场、主要干道和位于圣克拉拉工业区的货运火车站，因此除了重新改造展厅以外，项目还涉及适应性再利用和城市再建造。用无处不在的飞机、火车和汽车来进行交通创新的理念给设计团队带来了重要的灵感。仓库和办公楼都多年失修，所以在试探性地拆除后，Blitz 设计发现这栋小建筑的结构很坚固，可以被采用。考虑到原有仓库和办公楼相对粗糙，还有多学科设计实验室也有待考察，因此，设计风格有意避开了场地原先的工业历史。

最初，厨房式的设计理念，例如家庭中心和餐桌等关键元素不断浮现。照此，Blitz 设计将厨房作为设计理念，把所有用户聚集起来，同时还将它作为两栋建筑的主要接口。办公空间恰好设在中央，而进入建筑，首先出现在眼前的是专为会面与用餐而设立的工作咖啡区，它的酒店功能给客户和用户一种温馨和宾至如归的感觉。One Workplace 的工作人员可以在设在高处的会议室和观察台上快速地观察地面情况，为客户展示各种系统方案是如何相互交织并创出一个统一、灵活和分层的工作区布局的。

One Workplace 原办公室使用时间超过十年,新总部必须能够使用更长的时间。Blitz 设计通过在公共办公区中打造调升的楼层布置,来让办公室更灵活、寿命更长,除此之外,还有利于未来家具的重新配备,有限的色彩和图形也能更好地适应潮流的改变。

在更多地了解了 One Workplace 是如何与客户的合作之后,Blitz 设计发现环境也是影响销售的关键因素。客户将会购买适应整个空间和整个环境的家具,而非仅仅购买单件产品。因此,建筑需要衬托产品。个人空间的整体设计能让客户产生一种与空间的情感联系。

该项目不仅是一个创意设计，还产生了很高的经济效率。One Workplace 公司从面积约为 4 180m² 的空间搬迁到面积约为 3 252m² 的空间，然而职员数却从 101 人增加到了 165 人。这种效率的实现得益于工作台的减少，以及大多数的销售团队移到了移动办公区，不再需要固定办公桌。这些移动员工可以将他们的私人物品集中储存，工作时则可以到公共工作区，或是咖啡工作区，或者是交替工作区。One Workplace 所执行的正是一种现代工作模式。

PETZL North American Headquarters and Distribution Center

PETZL 北美总部和配销中心

Architects: ajc architects

Renderings, Elevations, Site Plans and Floor Plans: ajc architects

Interior Design: ajc architects

Client: PETZL Crolles, France

Location: West Valley City, UT, US

Photography: Alan Blakely

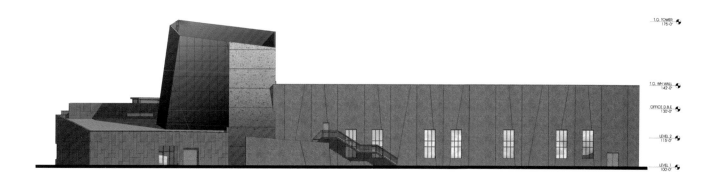

WEST ELEVATION

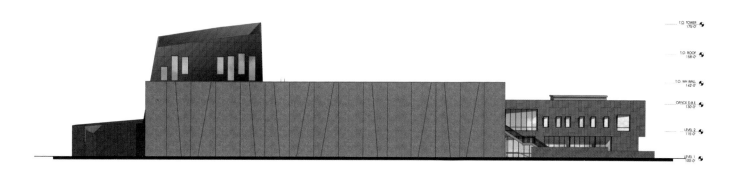

SOUTH ELEVATION

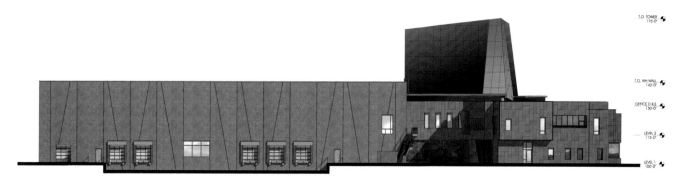

EAST ELEVATION

NORTH ELEVATION

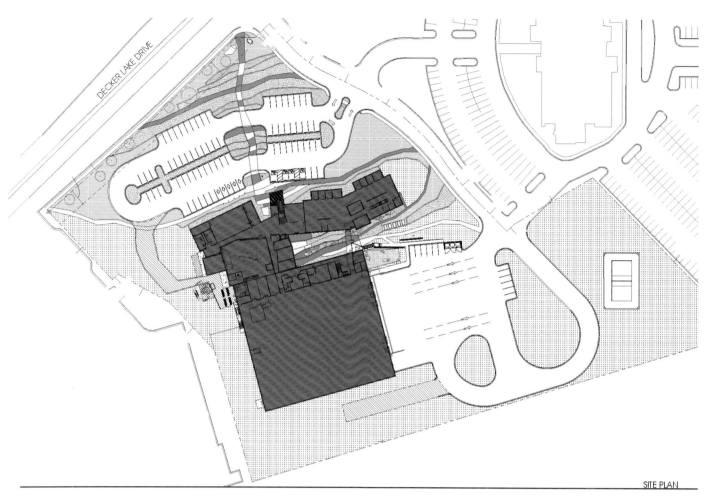

SITE PLAN

Petzl America recently completed their new North American Headquarters in Utah, which is on track for LEED platinum certification. Petzl's business is based on verticality and light, and the building is a metaphor for these concepts. The overall design incorporates a long list of sustainable features including a roof mounted photo voltaic array, natural daylight throughout all areas of the building including the distribution warehouse, bioswales on site to capture the rain water from the roof and re-charge the ground water, a plug-in for electric cars, recycled materials and products purchased within a 805 km radius of the building's geographical location.

The company has a dog-friendly policy that allows employees to bring their dogs to work, enhancing employee satisfaction for the work place. An exterior dog run was included in the site design, along with a community garden for employees with raised planting beds. The site is located adjacent to a TRAX stop, and the company encourages employees to use mass transit to work as well as bike to work. The building design includes a room for interior bike storage and a bike shop for employees to work on their bikes.

The amount of daylight available throughout the space, along with a strategically placed interior courtyard, minimizes the need for artificial lighting during daytime hours. All lights are controlled with occupancy sensors, and the entire building's energy use is monitored with a sophisticated Building Automated System (BAS).

Office spaces are designed with flexible office layouts with casual collaboration spaces scattered throughout, both interior and exterior. Low flow plumbing fixture was also selected to minimize the amount of water used throughout the building.

There is a 22m tall tower associated with the offices, and houses a 17m tall climbing wall – one of the tallest vertically in Utah. The training tower creates an exciting indoor environment, and has views at every level.

The warehouse portion has a high tech computer driven "Perfect Pic" inventory system – which is the hub of Petzl's North American product distribution system. This is another indication of Petzl's commitment to sustainability in making the work place more efficient, safe and enjoyable for the employees of the company.

Petzl had the vision to look at this building as more than "just a place to work", and through the company's values and respect for the environment and the desire to provide a healthy place for its workers and visitors, created a building that exemplifies sustainability for their North American Headquarters; a solid statement for the state of Utah and our commitment to sustainable building and design.

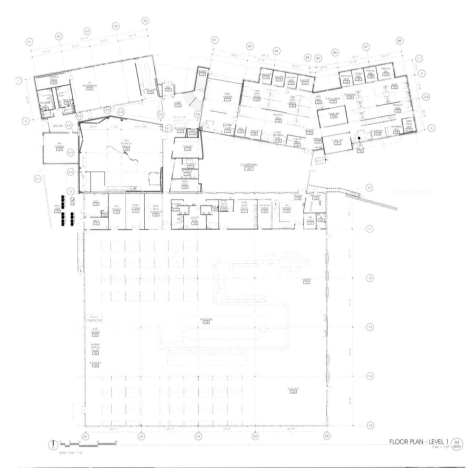

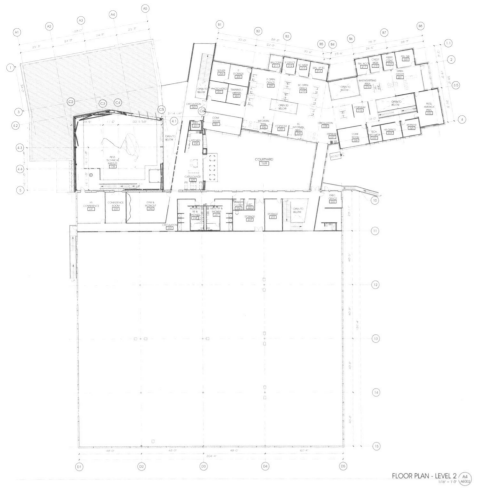

FLOOR PLAN - LEVEL 2

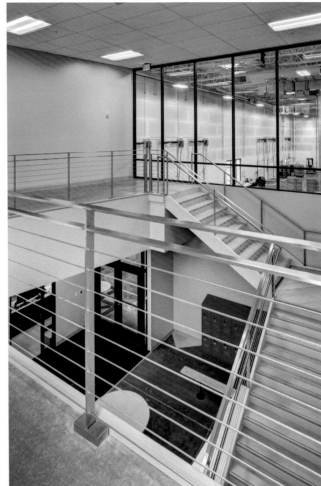

Petzl 美国公司在犹他州的北美总部最近终于完工，正在申请 LEED 白金认证。Petzl 商务是在灯光和垂直性的基础上建立起来的，而建筑恰好承载了这些理念。空间整体设计结合了各种可持续性设施，如屋顶上安装的光伏阵列。自然光能够渗透到建筑的所有角落，甚至是配送仓库。建筑所在场地的生态沼泽能够收集顺着屋顶流下来的雨水，再补给地表水。电动汽车充电设施、可再生材料和产品都是在建筑周围 805 km 范围内采购的。

公司倡导善待小狗，允许员工带着小狗来上班，这条政策提高了员工对工作环境的满意度。室外小狗活动区也纳入设计中，甚至还包括供员工种植花草的公共花园。场地靠近轻轨停靠站，交通便利。同时，公司还鼓励员工乘坐公共交通工具和骑自行车上班，因此设计了自行车停放室和自行车修理店。

空间的各个角落都有充足的日光，再配上设置巧妙的室内庭院，减少了白天时对灯光的需求。所有的灯光都可以通过灯控感应来控制，而且整栋建筑的能源也都通过先进的建筑自动化系统（BAS）来控制。

办公空间的布局十分开放、灵活，不管是室内还是室外，到处都是随意组合的空间。精心挑选的节水卫生洁具大大降低了整栋建筑所需的水量。

约 22m 高的高塔中设办公室和约 17m 高的攀岩墙，这道攀岩墙是犹他州最高的攀岩墙之一。训练塔中不仅有令人激动的室内环境，而且每一层都能观赏壮丽的风景。

仓库采用了完美的高科技来管理货物——电脑驱动库存系统，这就是 Petzl 北美产品分配系统的中心。同时，这也充分体现了 Petzl 在为员工打造高效、安全和轻松工作场所方面所付出的努力。

Petzl 总部不只是工作场所，更代表了公司的价值，公司对环境的尊重，以及公司为员工和客人提供健康的环境和创造可持续的北美总部方面所做的努力，充分证明了犹他州政府和 Petzl 在可持续建筑和设计方面的决心。

American Standard

美标办公室

Architectural Design: Space (Juan Carlos Baumgartner, Humberto Soto)

Participant Designer: Alejandro Danel

Lighting Design: Lua - Luz En Arquitectura

Location: Corporate Campus Coyoacán

Area: 1,500 m²

Construction: atxk

Furnishings: steelcase, Moblity

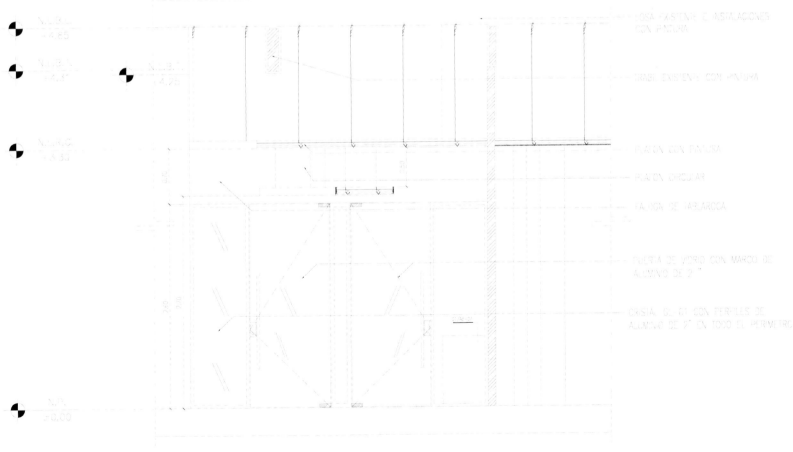

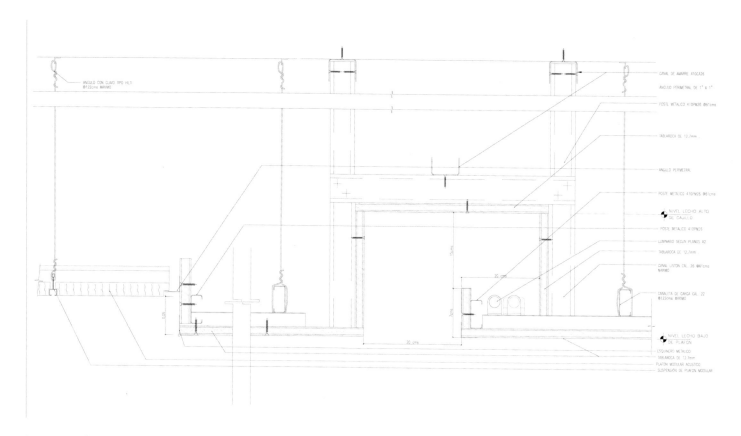

DETALLE DE CAJILLO LUMINOSO CON CAMBIO DE PLAFON LISO A MODULAR

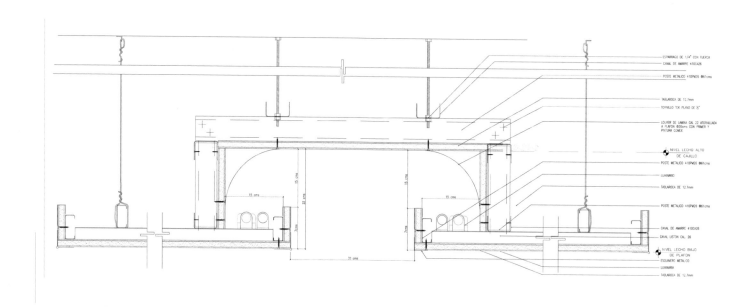

DETALLE DE PLATABANDAS SUSPENDIDAS CON CAJILLO LUMINOSO DOBLE

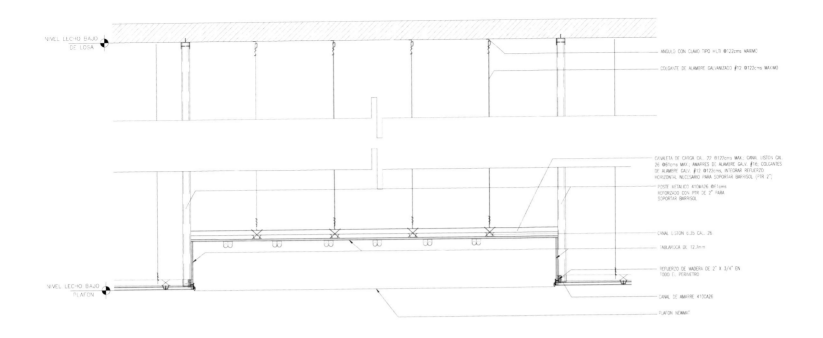

DETALLE PLAFON NEWMAT

The corporate offices of American Standard are located at the Corporate Headquarters, Corporate Campus Coyoacán, Av. Coyoacán 1622, México, D.F.

With a land area of 1,500m² the offices are conceived based on 2 central concepts which were fundamental for the development of the project. The first was to generate a showroom for its clients, and the second was the corporate offices.

The showroom was a singular element in the project, requiring a flexible space that could change easily, for which a lighting analysis was worked on in order to be able to have a space that, whatever the arrangement of the furniture, would be well lighted. The materials that were used for this space were carefully selected in order to generate a neutral background that would look good during the different furniture shows.

The panel design, subtle lines in the marble and rug cutting are based on the concept of "Water", curvilinear elements that simulate the waves that are transformed in panels that literally simulate the continuity in the space.

Access to the offices consists of a spacious reception with a waiting area, coffee area and 2 meeting rooms for clients. From this space there is access to the showroom or the offices.

As part of the architectural program, we have meeting rooms, a dining room, telephone booths, a training room, the general director's private office, and a very important space, the laboratory, where tests are done on the equipment and their different products in use are shown.

The offices also have men's and women's bathrooms. In these bathrooms it was sought to generate a homey touch that was different from those normally found in offices.

For the design of these offices, aspects were considered that altogether generate a much more comfortable space for the user, such as the use of natural light, lighting efficiency, standardization of work positions, more open work stations in order to promote communication among work groups.

美标总部办公室位于墨西哥联邦区科约阿坎大道1622号工业园内。

该办公室占地1500m²，被构思成两个部分：为客户打造的展厅和公司办公室，这两个部分共同组成了项目开发的基础。

展厅是项目中的特殊元素，它要求空间灵活，容易转变。此外，空间的灯光选择也下了一番功夫，不论内部家具如何摆放，空间都灯光通透。淡色背景是用精挑细选的材料打造而成，因此，不论是何种家具展，都能烘托展品的效果。

面板的设计、大理石的微妙线条以及地毯的形状都是以"水"的概念为基础的。而波浪般的曲线元素被应用到面板上，使空间更有整体感。

宽敞的接待厅中设有等候区、咖啡区和两间会见客户的会议室，由此空间能够直接通往展厅和办公空间。

会议室、餐厅、电话区、培训区和总经理办公室都是建筑项目的一部分；尤为重要的是实验室，各种设备都需在这里测试，还有各种在用产品也都展示在此空间中。

办公室中设有温馨舒适的男女卫生间，与通常的办公室卫生间完全不同。

在设计这些空间的过程中，各个方面都经过了深思熟虑，从而使空间更加舒适，比如充足的自然采光、标准的工作座位，以及促进团队协作的开放工作区。

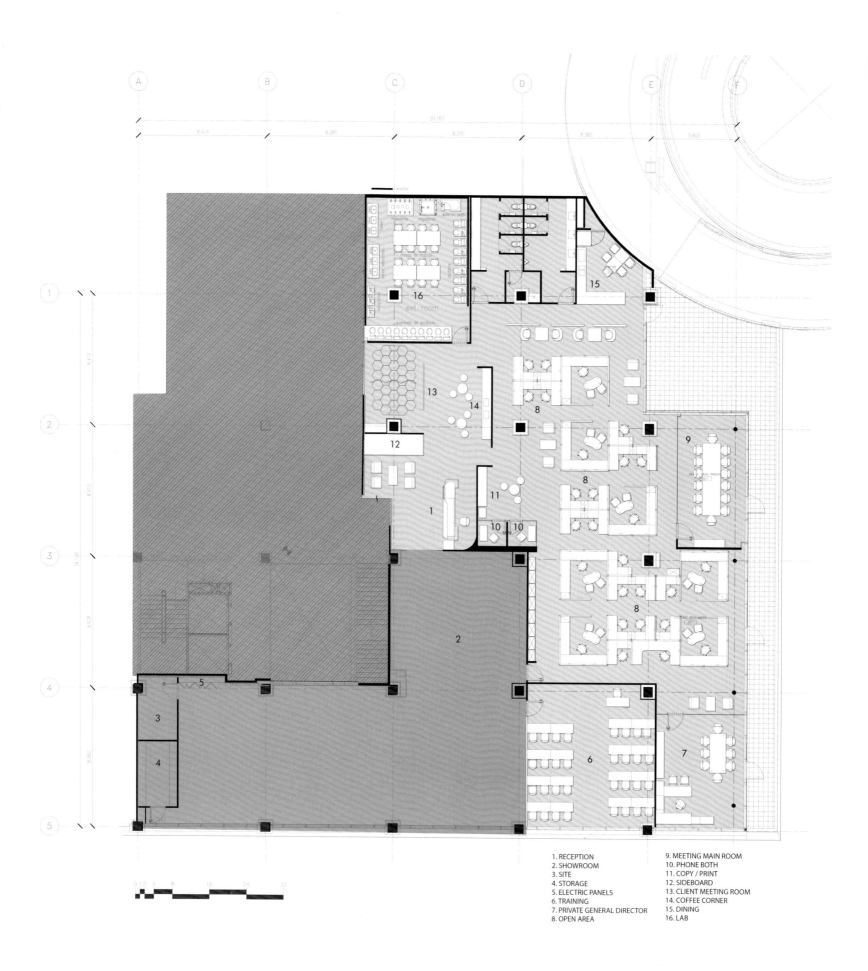

GAMING

Chartboost San Francisco Office

Chartboost 旧金山办公室

Architects: Design Blitz

Location: San Francisco, US

Area: 3,901 m²

Photography: Jasper Sanidad

When moving into a new office space, Chartboost was most interested in taking the opportunity to celebrate what is at the core of their business: video games. As the world's largest games-only technology platform, Chartboost wanted their San Francisco space to reflect the energy of the gaming world and create a fun, interactive environment that would excite anyone who entered. Blitz stepped up to the challenge and designed a modern office that brings the world of games to life.

The full Chartboost experience begins after walking up the stairs from entrance to the reception, where an entire wall is dedicated to the "photo op." Visitors can find their games on display and take pictures with the boxes (although we think the show-stealing, 3.4m tall dinosaur probably gets some photo love too). The dinosaur, aka "Buster," has become the Chartboost mascot - they even developed their first in-house mobile game starring the T-Rex. Hanging app boxes and an interactive iPad love wall further celebrate the developers and clients of Chartboost's products.

Blitz reflected the company's energy and focus on fun by using bright colors (limited to the 4 brand colors), playful textures, wall graphics, 3D sculptures, and a game room complete with ball pit and giant Lego wall. Adding an additional jolt of energy to the space are 5 "super themed" conference rooms that bring the classic games of Donkey Kong, Super Mario, Oregon Trail, Tetris, and Legend of Zelda to life. By combining custom graphics with specialty furniture and artwork, each room inspires the user to imagine they are inside the world of the game. These rooms have been so popular with the Chartboost team that additional ones are now in the works.

Chartboost's work style is super flexible - with teams changing and rearranging as frequently as every 2 weeks. Blitz designed a workstation layout that allows people to take their desks with them and gang together in a multitude of organic pods as needed. However, after 3pm most employees can be found lounging in one of the many themed open areas such as the Jurassic Park or the Mario Park. The latter is both a hall-hands area and a café, and has stadium seating for presentations and benches/tables for dining.

对于新办公室，Chartboost 最感兴趣的是让空间体现公司的核心业务——电子游戏。Chartboost 是世界最大的专攻游戏的科技平台。该公司希望新旧金山办公室能体现游戏世界的力量，具有有趣的互动氛围，能让任何一位进入到空间的人感到兴奋。Blitz 设计公司接受了这一挑战，设计了这个把游戏世界变为现实世界的现代办公室。

从门厅通往接待台的是一道楼梯，全面的 Chartboost 空间体验便开始于此。此处的整个墙面都十分适合拍照。访客可以发现自己的游戏被展示出来，也可以与这些盒子拍照（尽管我们认为高于 3.4m 的恐龙更引人注目，与它也能拍出喜爱的照片）。这只名为"巴斯特"的恐龙是 Chartboost 的吉祥物，公司开发的首个手机内置游戏就是"霸王龙"（T-Rex）。悬挂的应用盒以及交互式 iPad 爱心墙则是对 Chartboost 产品开发者和客户的赞美。

Blitz 采用了各种手法来体现公司的活力和趣味性，例如鲜艳的色彩（限制在 4 种品牌色彩）、有趣的材料、壁面图案、3D 雕塑和游戏室，还有让空间更加完善的球井和巨型乐高墙。5 间"超级主题"会议室给空间增添了活力，把经典的游戏带入到生活中，例如大金刚、超级玛丽、俄勒冈小道、俄罗斯方块以及塞尔达传说。通过把特制的图案与专门的家具、艺术品结合起来，给人一种在游戏世界中的感觉。这些空间在 Chartboost 职员中特别受欢迎，因此正在计划建额外的类似空间。

Chartboost 的工作方式十分灵活，团队队员更换和重组十分频繁，每两周一次。Blitz 设计的办公室布局允许职员根据需求携带着办公桌，在许多的独立小间中与组员一起工作。下午三点钟以后，大部门职员会在开放式主题空间中休息，例如侏罗纪公园和马里奥公园。马里奥公园还充当了走廊和咖啡区，并摆放了为展示准备的露天座位和用餐的长凳和桌子。

Lego Turkey

土耳其乐高办公空间

Architects: OSO Architecture

Design Team: Okan Bayik / Serhan Bayik / Ozan Bayik / Kerem Karatas / Aziz Emre Mert

Client: Lego

Location: Istanbul, Turkey

Area: 350 m²

Photography: OSO Architecture / Gurkan Akay

Lego Turkey is located in Mecidiyekoy, Istanbul with their offices within an area of 350 m². The location as the first office of the company located in Turkey has a staff consisting of 23 personnel. Highly "compact" and designed in accordance with an intense requirement program, the spaces have been generally planned in an open-office construction to relieve the intensity. In this sense, the reception and the small meeting room are located close to the entrance and at the center, while the open office areas are located next to these places. All of the manager and director offices have been planned to be along the facade due to their natural light and view advantage. The social areas in the office have been kept to a minimum because of the services provided by the building where the offices are located and abovementioned "compact" planning setup. In this sense 2 multipurpose "Loggias" have been designed. These spaces are intended for both socialization, and also the "informal" meetings in the office. In addition, there is a large meeting room on the right side of the entrance hall and at a directly accessible position. This meeting room for 14 persons is directly connected to a storage area that allows for brand product display.

As the world leader within their own field of industry, "Lego" brand and company have been recognized as primary design criteria for corporate identity highlight and interior design. In this sense, there has been inspired from the Lego products to strengthen the emphasis on the belonging to the brand identity of the designed spaces. In this respect, the entrance hall and the ceiling of the reception have been designed in "Lego Brick" form. In addition, "Lego Minifigures" have been used both on the walls and also glass partitions. The historical photographs of the company founder, the first employees of the manufacturing plant and the very first production stage have also been used at different places inside the office, and thereby, the intended sense of belonging has been consolidated. In general sense, Lego's corporate colors have been used on different surfaces and in different forms in all places, and a kind of harmony has been achieved in the interior design. In addition, acoustic comfort has been increased by taking special acoustic measures throughout the office. In this respect, high-density rockwool panels have been used for the open office ceilings. These panels have been integrated with luminaries thanks to the designed hanging and connection details and have been transformed into design elements. Moreover, double-glazing has been applied to all of the room partitions, and for all enclosed volumes - different from the open office - there have been used carpets, and thus, acoustic comfort has been increased.

In the office where a dynamic and modern design language has been adopted in general sense, it has been aimed to create venues which are integrated with the corporate identity, have high energy, and where comfort conditions are provided at a high level.

土耳其乐高办事处位于伊斯坦布尔梅西迪耶科伊，占地350m²，是该公司土耳其分公司的第一个办公区，职员人数为23人。由于空间很小，而且设计需要遵循严格的要求，因此空间采用了开放式的布局，以减少空间的拥挤感。接待区和会议室也因此设在了入口处的中央位置，紧接着就是开放式办公空间。所有的经理和主管人员办公室都设在靠墙的位置，自然光线充足，视野开阔。由于办公室所在建筑提供了一定的服务，而且空间采用了紧凑的布局规划，因此，社交空间占用的面积保持到了最小。按照这种理念，还设计了两个多功能的"凉廊"，可供社交和非正式会议使用。除此之外，门厅右边还设有大会议室，可容纳14人，进门便可直接通往这一空间，而且与展示品牌产品储存区相连。

作为业界的领军品牌，乐高将公司及其品牌作为主要室内设计标准来突出企业的形象。因此，空间设计从乐高产品中汲取灵感，从而赋予空间品牌归属感。照此，门厅和接待区的天花板被设计成"乐高积木"的形式。墙壁和玻璃隔墙上都有"乐高人仔"标志。公司创始人的历史照片，制造厂的第一批工人，以及最先的产品平台都被采用到空间的各个地方，从而增强了归属感。乐高的标志性色彩在不同的墙面以不同的形式展示出来，使室内空间更加和谐。除此以外，办公空间经过了特殊声音效果处理，让声音听起来更加舒服。而且开放式办公室的天花板采用了高密度石棉板，并嵌入了灯具和悬挂连接的细节物件，这些都成为了设计元素。不仅如此，空间内的隔墙以及封闭空间的墙壁都采用了双层玻璃，这与开放式办公空间不同。此处的地面铺上了地毯，使声音听起来更加舒适。

办公室充满活力和现代感，与企业的形象相契合，空间的活力和舒适度都达到了最大化。

Microsoft Gurgaon

吉尔冈微软办公室

Design Agency: DSP Design Associates Pvt. Ltd.

Design Leads: Geetika Jain & Shweta Grover

Design Team: Prithvi & Harvinder Singh

Location: Gurgaon, Haryana, India

Area: 6,689 m^2

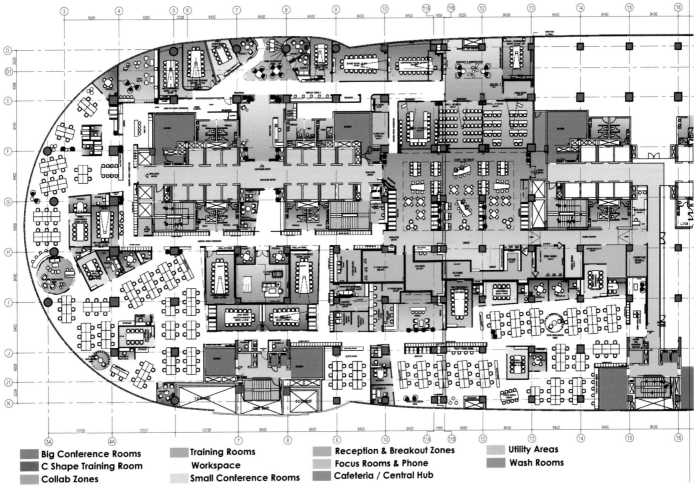

- Big Conference Rooms
- C Shape Training Room
- Collab Zones
- Training Rooms
- Workspace
- Small Conference Rooms
- Reception & Breakout Zones
- Focus Rooms & Phone
- Cafeteria / Central Hub
- Utility Areas
- Wash Rooms

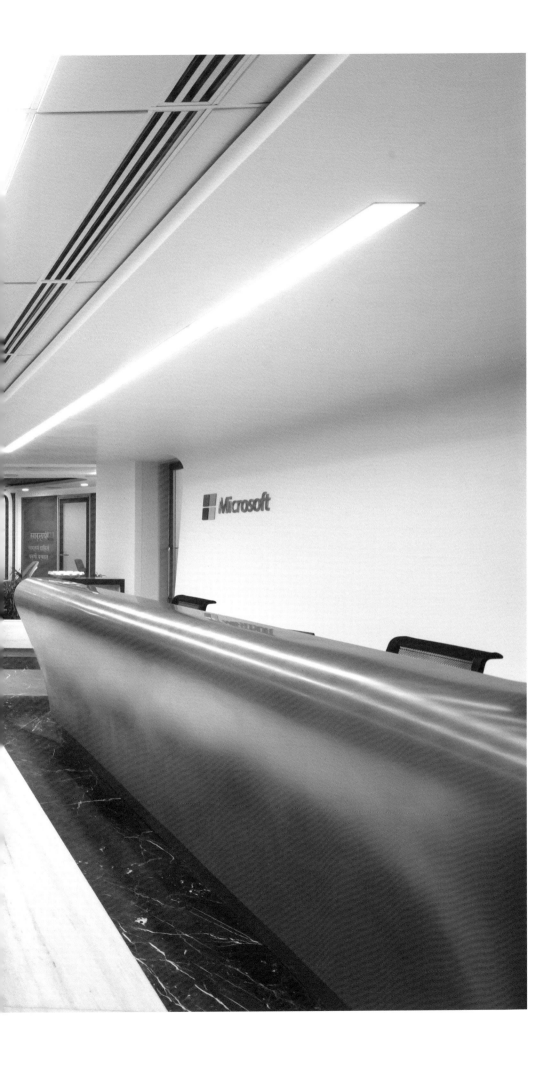

The one line brief given to DSP for Microsoft's Gurgaon office was "It should feel like a futuristic workspace with a sense of Indian ethos", very unique and beyond the ordinary. The global team at Microsoft was looking for ideas that would support chance run-ins with other employees, impromptu huddles, scrums, brainstorming sessions, work groups and formal meetings as well as breakout and relaxation zones.

DSP derived the inspiration of concept from the history of origin of city Gurgaon. It was home to Guru Dronacharya, the Great scholar of ancient India and the name of place "Gurgaon" is formed from Guru+Gaon.

This became the inception of the idea that Microsoft Gurgaon is a place where ideas evolve, where learning conceives, in a natural environment-like in the Indian tradition of the "Gurukul", a practice during the time of Guru Dronacharya.

And therefore a design ideology is adapted which depicts:
Nature – Element of outdoors brought indoors.
Fluidic – Free flow of forms & shape.
Learning – Depicted through color "Orange" color of intellect & graphics with 'Sanskrit' shlokas.
This ideology is spread through the interiors in a subtle way.

The new workspace, built along Microsoft's Global Workplace Advantage Program guidelines, is kitted out with 60 meeting rooms of varying sizes and specifications, from phone rooms for one-on-one conversations, focus rooms that can accommodate 4 – 6 people at a time, to meeting rooms for larger consultations. For all-hands meetings, retractable walls in the cafeteria can be pushed out to merge with the training rooms and create a town hall. Variety of spaces for Spontaneous collaboration, brainstorming and teamwork are key tenets of the new office.

The facility is designed with a sharing ratio of 2.4:1 (2.4 seats on average per employee) and all the other spaces apart from workstations are also considered work seats. So all the collaboration spaces, café, lounge, focus and phone rooms are designed in such a way that people can work there for longer hours. Free seating and flexi timings, high-speed wireless connectivity and plug points in all parts of the office, seating options ranging from sofas to hi stools and office chairs, and furniture – including tables – that can be moved around for impromptu meetings, are all ideas that emerged from the need for greater mobility and an "activity-oriented" office.

Although the office space is designed to be an unsigned and highly shared workplace, yet keeping in the personalisation of the workplace by the end users, lockers have been provided which is the only personalised storage space for the employees.

For individual, heads-down work at the office, there are quiet zones. The Lynch app, is used widely in the by people to publish their location so others know exactly where they are in the office – this is useful in the absence of assigned seats this also allows employees to book meeting rooms remotely.

Microsoft's Gurgaon office has taken technical evolution to a level leaps and bounds ahead of the contemporary Indian office scenario.

在设计吉尔冈微软办公室时，DSP 设计公司仅仅收到了一条简短的要求，也就是"办公室应该是具有印度韵味的未来派工作空间"，这显然非常独特。微软全球团队寻求的是一个能支持各项活动的空间，例如，职员之间的交流与竞争，临时小型会议和"头脑风暴"会议，集体工作与正式会议，以及休息和放松空间。

DSP 从吉尔冈城的历史渊源中寻找灵感，得知吉尔冈的名称"Gurgaon"是由"Guru+Gaon"组成，还是古印度伟大学者 Guru Dronacharya 的家乡。

这便是吉尔冈微软办公室设计灵感的雏形：这是一个想法和学问在自然环境中形成的地方，就像是印度传统中 Guru Dronacharya 时代的"学院"。

因此，设计理念所表现的是：
自然——把户外元素引入到室内；
流畅——形式和形状的自由转化；
学问——代表智慧的橙色和梵文图形。
这种理念在室内空间微妙地展现出来。

新工作空间是按照微软全球优化工作环境（WPA）项目指导方针设计而成，设有 60 间大小、功能各不相同的会议室，例如一对一交谈的通话间，一次可以容纳 4~6 人的专题解决工作间，以及大型咨询会专用会议室。自助餐厅中的可拆除墙被撤掉后，便与培训室组成可以举行全体会议的大厅。

工作区的设施是按 2.4:1 的分配比率（2.4 座位 / 人）设计的，而与工作区分开的其他空间也设置了工作座位。因此，人们可以在所有的合作区、咖啡区、休息区、专题解决工作间和通话间中工作更长的时间。办公室的各个角落都有自由座位、自由活动区、高速无线网和联网端口；座位的种类也各式各样，包括沙发、高脚凳和办公椅等；家具也都是可以搬动的，甚至连桌子都可以搬去开临时会议。所有的这些想法都满足了对更大的流动性和活动型办公室的需求。

尽管空间被设计成没有分区、高度共享的办公室，然而为了保证工作空间使用者的个性，便在空间内安装了储物柜，这也是为职员设计的唯一的个性化储物空间。

安静区可供个人专心埋头工作。"林奇"定位应用软件可以用来告知别人自己的确切位置，在职员之间很受欢迎。这个软件对于一个没有固定工作位的办公室来说十分实用，职员还能用它来远程预订会议室。

微软吉尔冈办公室在技术革新上有着飞跃性的进步，在印度现代办公室设计中遥遥领先。

•181•

HARDWARE/SOFTWARE DEVELOPMENT

Microsoft – Redmond Building 44 Offices

微软公司——雷德蒙 44 号办公楼

Design Agency: ZGF Architects LLP

Client: Microsoft

Location: Washington, United States

Area: 14,678 m²

LEVEL 1

- Conference Room
- Core/Support
- Focus Room
- Hub: Kitchen/Lounge
- Occupant Specialty Space
- Open Meeting Space
- Open Office Neighborhood
- Phone Room
- Reception
- Team/Work Room

The 14,679 m² 3-storey workplace redesign is a renovation and tenant improvement of Microsoft's Building 44, an existing 1980s era building located on Microsoft's main campus in Redmond, Washington. The programming and design concept realizes Microsoft's Workplace Advantage (WPA) 2.0 program, aimed at providing the best workplace environment to meet the needs of employees and drive business forward. Microsoft conducted research and found that employees were spending less time in individual offices. To support the company's workforce shift, the design incorporates a variety of collaboration space while decreasing individual space.

The design team worked closely with the users to fine-tune the size and functional requirements for the workspace neighborhood concept and create a flexible environment for evolving tenant needs. A chief driver for the re-design was to enable agility and speed to market by removing barriers between people and providing choices in workspaces. The design provides a balance of spaces that support all types of work, from formal to casual, and from team to individual while celebrating individual achievements and fostering collective community to provide the tools needed for collaboration and innovation.

该办公楼分为三层，面积约为14 679 m²，坐落在华盛顿雷德蒙微软主园区中，建于20世纪80年代。本项目是将这栋微软44号楼进行翻新设计。办公楼的规划和设计理念都符合了微软优化工作项目2.0（WPA 2.0）的要求，旨在营造一种能满足职员需求和推动业务发展的最佳办公环境。微软对此展开了调查，并发现职员在个人办公空间中花费的时间其实很少。为了实现办公室工作区流动使用的模式，设计中加入了各种合作空间，并减少了独立工作空间。

设计团队与空间用户密切协作来调整空间大小和功能要求，以实现"工作区邻域"的理念，同时创造灵活的环境以满足不断增长的企业需求。重新设计的主要动力是通过移除人们之间的障碍，以及在工作空间中提供各种选择，让对市场的灵敏性成为主导。这种设计实现了空间之间的平衡，使空间能够适应从正式到休闲式，从团体式到独立式等各种工作模式，在赞扬个人成就的同时，也鼓励团队合作，为合作和创新提供各种需要的设施。

LEVEL 2

LEVEL 3

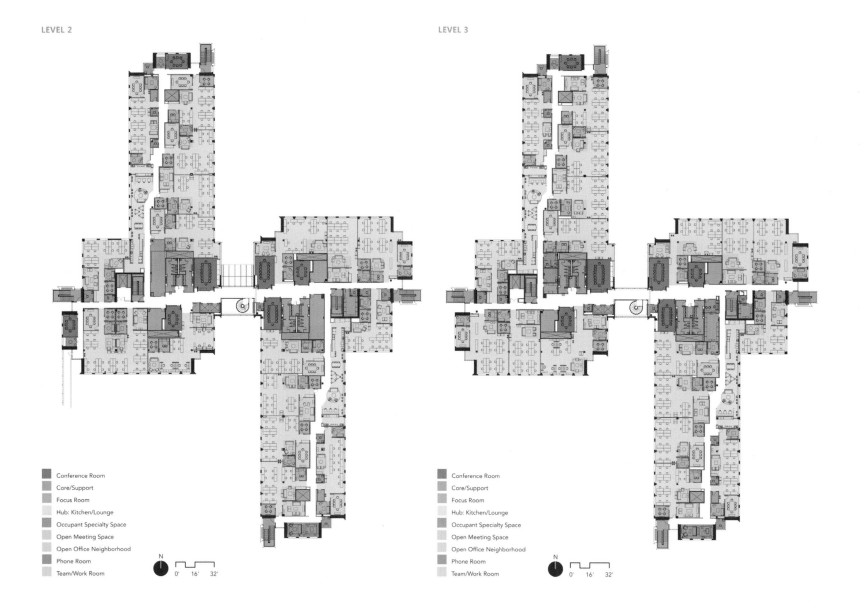

- Conference Room
- Core/Support
- Focus Room
- Hub: Kitchen/Lounge
- Occupant Specialty Space
- Open Meeting Space
- Open Office Neighborhood
- Phone Room
- Team/Work Room

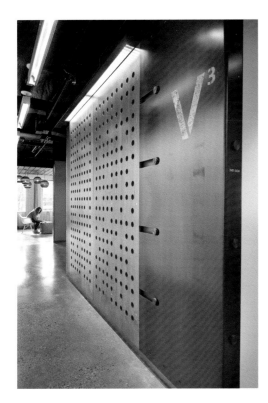

HARDWARE/SOFTWARE DEVELOPMENT

Meltwater Tenant Improvement

融文集团办公室

Architects: Design Blitz

Location: San Francisco, United States

Area: 2,415 m²

Photography: Jasper Sanidad

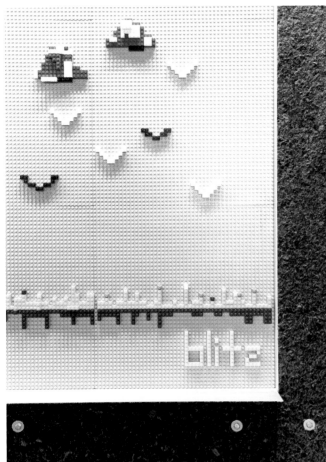

Meltwater, a leading provider of online intelligence solutions, was founded in Norway in 2001 and has 57 offices worldwide. To accommodate its growing staff, the company relocated its San Francisco world headquarters to the 10th floor of the historic Standard Oil building on Bush Street. Meltwater wanted a space that would celebrate its heritage and strong work ethic, while keeping true to the lively setting of San Francisco. The company partnered with Design Blitz to create a clean, contemporary office that combines classic, Scandinavian modern design with cutting-edge, San Francisco tech industry style.

Once the programmatic adjacency layout was formalized, Design Blitz looked to the company's brand identity and core values for design direction, starting with the description of the company name:

"Meltwater – the transformation of ice and snow to crystal clear, flowing water, is a symbol of renewal and change."

Using the idea of water as both a circulatory and transformative element, Blitz designed the space with Norway's largest river as a reference. The Glama River, which runs from icy mountains in the north through farmland and into a small urban center in the south, flows past many different topographies and ecospheres in Norway. Similarly, the Meltwater office has a main circulation path that runs through the office and all of the departmental zones. Blitz selected 4 ecospheres found along the Glama River – ice caps, waterfalls, green valley, and pond – to inform the design of the distinct zones in the Meltwater office.

Design Blitz specified colors, forms, and materiality in each zone to reflect its respective Norwegian environment. The reception features an articulated enclosure that is inspired by ice caps, with geometric

融文集团于2001年在挪威成立，作为一家领先的在线智能方案提供公司，已在全球建立了57个办事处。为了满足职员增长的需求，公司将其旧金山的总部搬至了布什街著名的"标准石油大楼"的第十层。融文集团希望新空间能体现公司的传统和强烈的工作责任感，当然也要符合旧金山的活泼氛围。为此，公司与Blitz设计公司密切合作，以创造一个清新的当代办公空间，同时融入经典的斯堪的纳维亚现代设计和先进的旧金山科技行业风格。

Blitz确定好大致的布局之后，便开始研究公司品牌特征和核心价值，以确定设计方向。项目的起点便是对公司名称的描述：

"融水（Meltwater）——冰雪融化成清澈的流水——象征着复兴和改变。"

hanging light fixtures evocative of ice crystals and a custom, angular geometric desk that references layered, sheathed ice. For the area around the work cafe, Blitz placed decorative, custom-patterned screens representing the flowing water of a waterfall to create a vibrant backdrop for social activities. The Engineering zone in the western wing of the office uses a pixelated, bold green carpet pattern and multilevel ceiling panels to depict the green tiles of farmland in Norway and to brighten up a previously light-deprived space. Lastly, the Central Operations zone in the office's eastern wing incorporates cool blues and a custom lighting element symbolizing water droplets to achieve the serenity of a calm pond.

Along with the transformative water motif, Blitz called upon the heritage of Scandinavian architecture to produce an overall design that would unify the space and evoke the company's history, while keeping things true to the technologically advanced setting of San Francisco. Within the distinct design schemes and throughout the office, a unified palette of materials consisting of warm woods and vibrant, crisp colors was specified not only to accentuate Meltwater's heritage, but also to update it with a current, San Francisco style. Another key aspect of the overall design is the transparency of private offices to allow natural light to flow and increase collaboration. The architects used glazing at all the offices wrapped in wood overhangs to encourage the flow of natural light. Cutouts in walls form an abstraction of the Norwegian flag, tying into Meltwater's roots in a playful way while also evoking a classic, Scandinavian modern form. Additionally, Blitz added a Lego wall to an otherwise unused leftover space to add a sense of levity to the project and to encourage an alternative to a traditional meeting space. Finally, the inclusion of touch screen monitors, state-of-the-art audio systems, and high-definition projectors solidify the project's place in the tech hub of San Francisco and support a company that is constantly adapting.

The result is clean, dynamic space that respects Meltwater's heritage while looking to the future, providing a home for the company as it continues to grow and evolve.

水是一种不断循环和转化的元素，Blitz便采用了水的理念，并参照挪威最大的河流来设计空间。格洛马河从北边的冰山上流淌下来，穿过农田，流到南边的小城市中央，它流经了挪威不同的地形区和生态区。同样，融文办公室也有一条从办公室到各部门区域的主流通线。Blitz从格洛马河流经的生态区中挑选了4种元素，包括冰盖、瀑布、山谷和池塘，来区分办公室中的各个空间。

每个分区中的用色、形式和取材都经过了精心的设计，以便显示独特的挪威环境。接待区的设计是受冰盖的启发，呈现出铰接式围墙结构，还有冰晶似的几何吊灯，以及层叠冰盖状棱角鲜明的几何桌子。咖啡区周围装饰着定制的屏风，上方的图案是一泻千里的瀑布，同时也为社交活动提供了颇有活力的背景幕。西翼的工程区大胆采用了像素化绿色图案的地毯和多层天花板，意指绿油油的挪威农田，也让稍显昏暗的空间瞬间明亮起来。最后是东翼中央的运营区，其中的安静蓝色和定制灯具象征着水滴，使空间具有池水般的平静。

除了富于转化的水以外，Blitz的设计继承了斯堪的纳维亚建筑的传统，赋予了空间整体性，也让人联想到公司的历史，此外，还与旧金山科技进步的背景相契合。统一的用材方案将不同设计方案的空间组合成一个整体，例如暖色木。鲜亮的色彩不仅突出了公司的传统，还让传统与时俱进，形成旧金山风格。整个设计的另一个关键点就是私人办公室的透明度，这不仅有利于自然光照入空间，更有利于促进合作。建筑师在所有的办公区都装上玻璃，并用悬挂的木板把玻璃包裹起来，从而让自然光自由地照入空间。墙壁上刻有抽象的挪威旗帜，将新办公室与融文的根基巧妙地系在一起，具有经典的北欧现代风格。除此以外，Blitz在多余的空间内加入了一道乐高墙，既给空间增添了一种多变感，也鼓励改变传统的会议空间。触屏显示器、最新的音响系统和高清投影仪让项目更加完善，巩固了该项目在旧金山科技中心的地位。另外，空间还能不断满足公司的发展需求。

这个清晰的动态空间继承了融文集团的传统，也充满了对未来的展望，是不断发展和改革的融文家族的家园。

HARDWARE/SOFTWARE DEVELOPMENT

Autodesk Israel – Tel Aviv

以色列特拉维夫市欧特克开发中心

Design Agency: Setter Architects

Designer: Shirli Zamir

Project Manager: Chen Yaron (Yaron- Levy)

Photography: Uzi Porat

Setter Architects has recently completed a new cutting edge development center in Tel Aviv for Autodesk, an international leader in 2D and 3D design software. The new office extends over 4 floors of a new tower on Tel Aviv's prestigious Rothschild Boulevard. Setter Architects was tasked with designing a space to reflect a creative blend of Autodesk's corporate and local cultures, which puts a strong emphasis on the well-being of its employees – even to the point of allowing staff to bring their dogs to work! With a planning directive to bring pleasure and fun to a flexible work environment, Setter was also tasked with creating a design that would highlight Autodesk's culture of collaborative teamwork, open communication between teams, and creating spaces that would work for both alone and in groups. Autodesk defines itself not just as a software company but also as a "design" company and ascribes tremendous importance to the appearance and functionality of its offices. The project team embarked on an engaging process of capturing the needs and expectations of both local managers and executives from the company headquarters in San Francisco. This dialog resulted in the use of rich three-dimensional design and a broad range of materials, applying the language of software programs that Autodesk develops in its design and graphics, and creating a truly pleasurable work setting.

A special feature of the new development center is to take advantage of the magnificent 360° surrounding city view of the beautiful modern Tel Aviv. With that in mind, the space was designed for all employees to enjoy the views from every corner of the office space. The interior design also includes an assortment of simple and raw industrial-style materials, from concrete floors to rusted treated metal walls, reclaimed wood from discarded window frames and plaster, and vibrant wallpaper with graphic elements infused with Autodesk branding. The intriguing blend of these elements creates a unique sense of flow.

As befits a company professionally engaged in 3D, a sense of a three-dimensional space is incorporated into the design of the masterfully angled conference room and an interconnecting stairwell with visually striking three-dimensional concrete wall.

The office plan offers ample quiet work-areas, informal collaboration areas, and formal conference rooms, which also serve as visual and acoustic barriers in the transition from public spaces and work areas. These rooms are easily accessible from the work areas and the public spaces, to maximize employees' options on how

they would like to work and meet.

Sustainability at Autodesk is a company-wide initiative involving all aspects of Autodesk's business. The project was planned and designed in accordance to the guidelines of U.S. Green Building Council. The office has a sophisticated monitoring system controlling the lights, drapes, and air-conditioning, which continually monitors and tests the air quality in most enclosed rooms. The project is registered under LEED® Green Building Certification program pursuing the highest rating of LEED® Platinum in Commercial Interiors.

最近，Setter Architects 建筑公司完成了特拉维夫市欧特克新开发中心。欧特克是国际领先的 2D 和 3D 设计软件开发公司。该办公室位于特拉维夫市罗思柴尔德大道的新大楼中，分为四层。Setter Architects 的任务是设计一个融合欧特克企业文化以及当地文化的办公空间，突出让职员受益的空间特征，甚至包括允许职员带狗工作的特色。项目规划旨在营造愉悦身心、趣味十足的灵活工作环境。除此以外，突出企业的团队交流合作，创造单独工作和团队合作等多样空间，也是 Setter 的目标。欧特克公司不仅把自己定义为一家软件公司，而且还是一家设计公司，并特别强调办公室的外观和功能应当体现这两点。项目负责团队便开始收集本部管理者和旧金山总部高管们的要求和建议。在这些交流中，便形成了以立体设计形式、多样的材料以及图形和设计所用的软件程序语言，来创造真正富有乐趣的工作空间的设计方案。

新开发中心的特色就是拥有 360°特拉维夫市现代城市风光。为此，设计目标就是让所有职员能在办公室的任何一角欣赏到壮丽的风景。当然，室内设计中也采用了朴素自然的工业风材料，例如混凝土地面、锈化金属墙、回收再利用的窗框木材和石膏以及带有欧特克标志的鲜艳图纹墙纸。这些元素的有趣搭配形成了一种非凡的流

动感。

由于公司专门从事3D行业,因此三维空间的想法也被采用到了设计中,例如角度处理到位的会议室和三维混凝土墙楼梯井,两者都能给人以视觉上的冲击。

办公室的布局合理,设有充足的安静工作区、非正式合作区和正式会议室,这些空间还充当了公共空间与工作空间之间的过渡区,有效地维护了空间的隐秘性。而且要从公共空间和工作空间进入到这些区域中也十分便捷,给职员工作和会面提供了多样的场所选择。

可持续性是整个欧特克企业的精神,也深入到了欧特克公司业务的各个方面。本项目的规划和设计都遵循了美国绿色建筑委员会制定的指导方针。办公室中设有高级的监控系统,控制着灯光、窗帘和空调。针对封闭的空间,还能监测空气质量。本项目申请了LEED绿色建筑认证,获得了最高等级商业空间设计LEED金牌认证。

Headquarter of Microsoft in Lisbon

里斯本微软总部

Architects: 3g office

Location: Lisbon, Portugal

Photography: Filipe Pombo/Microsoft

Microsoft has recently moved into a new 4 floor, 6,317 m² flexible office facility in Lisbon's Park of the Nations. Designed by 3g office, the office makes use of technology in an effort to increase mobility, productivity, and collaboration.

The new office houses 475 employees and abandons the traditional concept of assigned workstations and instead offers a diverse landscape of work areas: workstations, phone booths, collaborative tables, focus rooms, brainstorming think tanks, and a variety of meeting rooms of differing sizes.

Together, this array gives employees access to privacy and openness. The designers note some rationale behind the open space: "Open workplaces encourage relations between employees. They are close to building fronts so as to ensure greater light flow and better illumination of the workspace. To encourage the spirit of collaboration, for tables we ran away from secretary type ones and looked for others with round, circular and oval shapes of different sizes.

This type of spaces eliminates the concept of territoriality. The only space that employees owned is a drawer of 0.10 m³ size. This is a place where employee creativity is allowed to identify that "territory".

One innovative solution was used to achieve both transparency and privacy without adding more rooms: "In the meeting rooms the choice made was to put glass "SmartGlass" which allow the user to choose the level of privacy he or she needs. With this type of glass we also ensure a maximum transparency between the spaces, what is in line with one of the key concepts of Microsoft philosophy: Life without walls."

In addition to housing employee workspaces, the Lisbon office has a more client-facing space designed, so Microsoft can promote its products and in a way that visitors can come and experience the space and Microsoft technology at the same time.

"The client areas are located in both the ground and third floors. We find a high diversity of spaces depending on needs of use, from a meeting to the organization of an event, with support spaces like a bar or the terraces in the three upper floors, so as to allow enjoying the fantastic surroundings of the Park of the Nation and the river Tajo."

The project also employed several non-traditional material choices in using locally sourced fabric and cork. The Burel fabric was used on several walls throughout which not only provides acoustic dampening qualities, but creates a memorable aesthetic to the rooms in which it was used. Cork was used as a flooring solution. The project also made use of Portuguese-designed lamps in both the front entry and larger meeting rooms.

微软公司葡萄牙总办事处最近搬至了位于里斯本国家公园的一个全新的办公空间中。该办公室分为4层，面积约为6 317m²，由3g工作室设计而成。科学技术的使用促进了人员的流通、协作和工作效率。

新办公室能容纳475位职员，摒弃了传统的分配式工作位，取而代之以多样化的工作区，其中包括工作位、电话间、协作桌、独立工作间、"头脑风暴"间，以及大小不一的各种会议室。

因此，职员拥有充足的私人空间和开放空间。设计师还解释了开放空间背后的原理："开放空间都设在楼层的边缘，空间光线充足。同时，我们并未选择秘书办公桌，取而代之以形状各异的桌子，例如，圆形、环形和椭圆形等，从而有利于加强职员之间的联系，激发合作精神"。

这种空间布局消除了领域性的概念。而职员所拥有的只是0.10 m³的空间，也只有在这里，职员的创造力才会区分"领域"。

本项目的创意之举就是在不添加房间的条件下保证空间的透明度和隐私性："会议室中的'智能玻璃'允许职员根据自己的需求来调节隐私度。这种玻璃还有利于保证空间与空间之间的透明度，恰好符合微软'生活无障碍'的理念。"

除了职员的工作空间以外，里斯本办公室还设有面向客户的空间，这不仅有利于微软推广产品，还有利于访客体验微软空间和科技产品。

"一楼和三楼都设有客户空间。多样化的空间都是根据用途设计的，不管是开会还是举办活动，都有像三楼的酒吧和露台等空间来支持活动。同时，人们还能欣赏国家公园和塔霍河的迷人风光"。

本项目采用了几种非传统材料，例如，当地采购的织物和软木。几面墙壁上的布雷尔布不仅能消减噪音，还给空间增加了美感。软木则是被用作地板。此外，门厅和大会议室中都安装了葡萄牙灯具，使空间显得富丽堂皇。

Unilever Brand Hub Europe

联合利华欧洲品牌中心

Architects: Fokkema & Partners Architecten

Area: 14,000 m²

Photography: Horizon Photoworks, Rotterdam

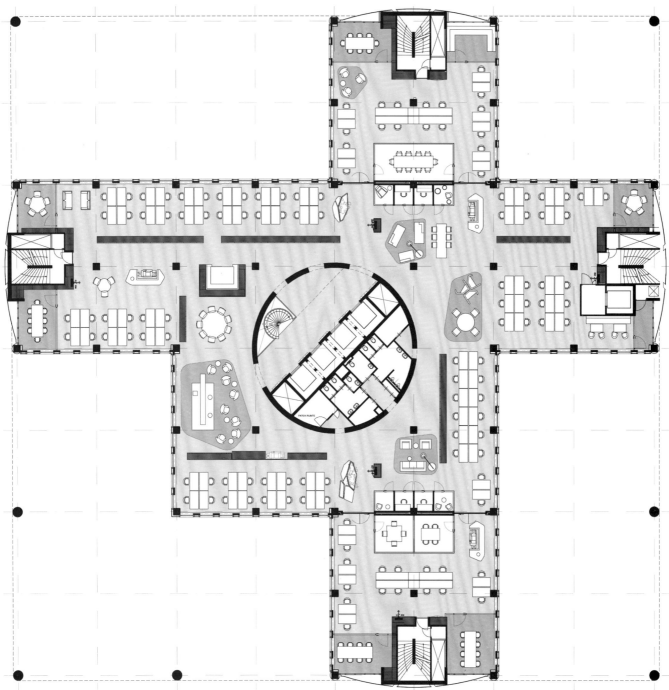

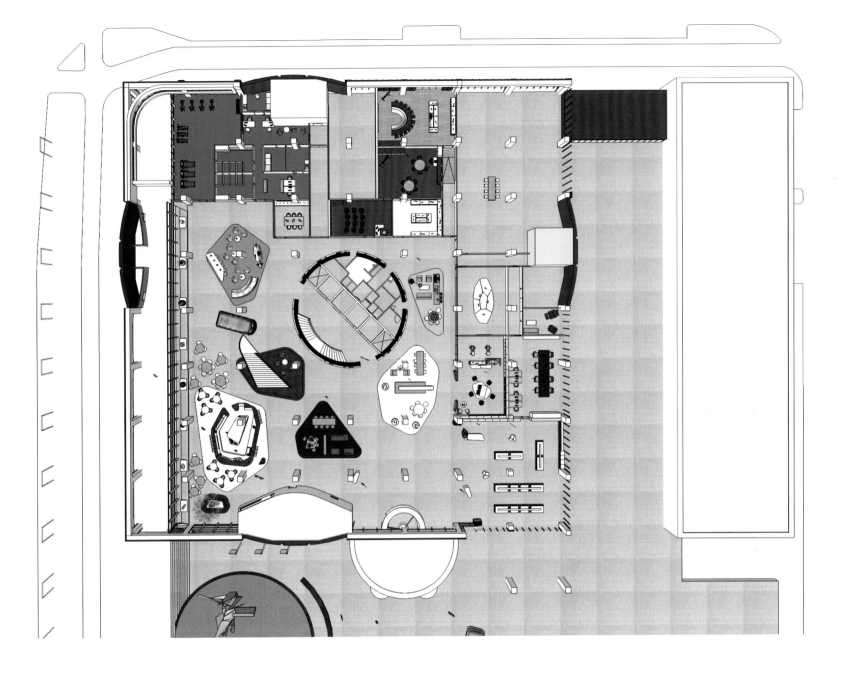

Fokkema & Partners Architecten have developed a new Head Office for Unilever located in Rotterdam, Netherlands. The new office is meant to the brand's European Hub.

At the end of 2012 Unilever decided to center all European Category Marketing Teams at the Head Office at Weena in Rotterdam, creating a true "Brand Hub Europe". The renewed Unilever Head Office emanates openness and liveliness, enthusiasm for the job and a strong but subtle appreciation of real stories and real life connected with every day products.

The ground and first floor are characterized by the "islands in the stream" where one can meet, work, interact, get informed, consume and present the jewels of Unilever, for both colleagues and guests. A new and inviting staircase, glass railings and new, floating floors in the atrium improve the connection between ground and first floor. Moving freely between islands, such as the Lipton Bar, Ben & Jerry's island, Personal Care area or Axe Bar, which each have a different function and atmosphere, the world of Unilever with its wide range of products comes to life.

The dark blue core reflects the overall corporate identity of Unilever on each floor throughout the building. Shelves enable employees to display the full bandwidth of the Unilever products. The studio-like work floors are organized around this product core. A natural background of wood, white, warm grey and skin tones form the basis in which the specific design and graphic elements can come to their full right. Colourful huddle islands in the open office area connect to the various meeting rooms made of full height glass. Together with a category experience point, each floor has become a true brand hub.

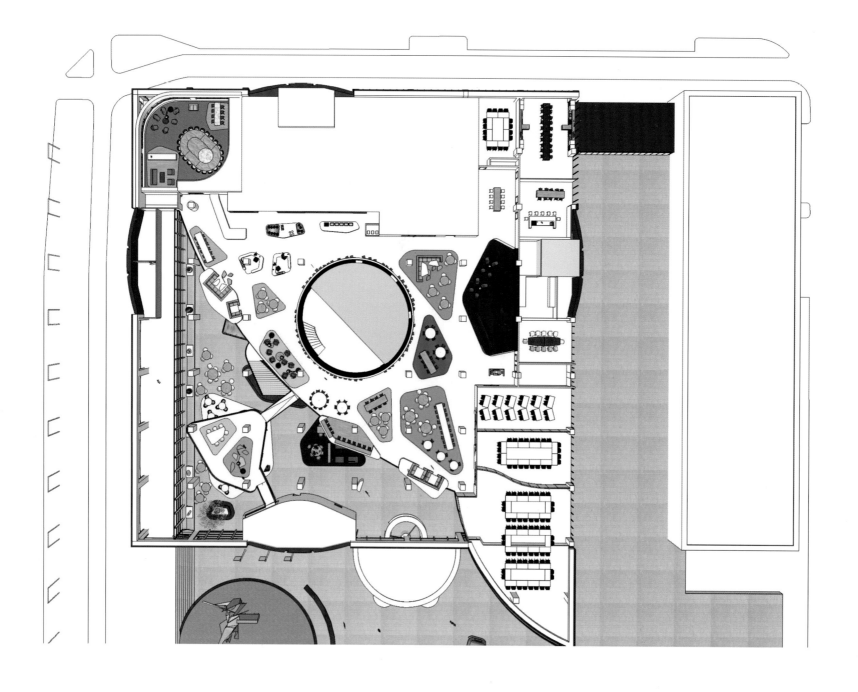

该联合利华新总部位于荷兰鹿特丹,由 Fokkema & Partners Architecten 设计完成,注定将成为该品牌在欧洲的品牌中心。

2012 年底,联合利华决定将其欧洲所有的销售团队都安置在鹿特丹威纳总部办公室中,创立一个"欧洲品牌中心"。全新的联合利华总部以开放和活力为特色,不仅能激发人们对工作的热情,而且每天与产品发生的真实故事,以及与产品接触的真实生活,会让人产生一种强烈而微妙的愉悦感。

建筑的一层和二层以"河中小洲"为特色,人们可以在此会面、工作、交流和休息,还能获取信息,甚至是给同事和客户展示联合利华的珠宝。新楼梯两侧是玻璃扶手,让人不禁想踏上楼梯。而中庭的悬浮楼梯巧妙地将一楼和二楼连接起来。在各"小洲"之间自由移动时,联合利华的世界和琳琅满目的产品便会一一呈现在眼前。这些"小洲"包括立顿茶吧、本·杰里小吃台、个人护理区和艾克斯吧。

每层楼的中央结构都是深蓝色的,代表着联合利华的整体企业形象,固定在此结构上的架子可供展示联合利华所有品牌的产品。工作区域都围绕着产品展示中心结构布置,就像画室一样。各区域都以木色、白色、暖灰色和肤色为背景,十分自然,烘托了设计细节和图形元素。开放式的办公区中设有色彩各异的"小洲",它们相拥相簇,将被全高玻璃包裹起来的各类会议室连为一体。总而言之,每层楼都是真正的品牌中心。

MANUFACTURING

NUON Amsterdam

阿姆斯特丹 NUON 总部

Architects: Heyligers design+projects

Area: 27, 000 m²

Photography: Rick Geenjaar

INSIDE/OUTSIDE OFFICE DESIGN V

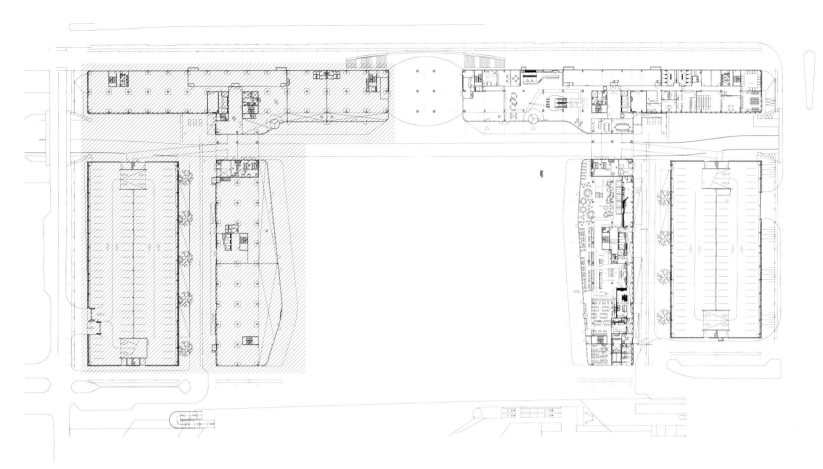

27,000 m² sustainable reuse and regeneration for "the new way of working" in Amsterdam Southeast.

The "nieuwAmsterdam" building was built 25 years ago and recently completely transformed into a modern mixed-use building with hotel, catering industry and 27,000 m² of office space for power company NUON.

In September of 2012, HEYLIGERS d+p was awarded the contract to provide the interior design for NUON and only 15 months later the new office was taken into use. With its new head office NUON embraced "the new way of working", work shifting: a fully open and flexible working environment.

The interior design consists of open work floors that are focused on diverse and flexible working by being interspersed with conference rooms, telephone booths, concentration workplaces, pantry's, "executive suites" and project rooms. The building offers diverse facilities for its users, including a reception lobby with a water bar, a restaurant with 650 seats, a library, an espresso bar, a service centre, a conference centre, and a sky lounge.

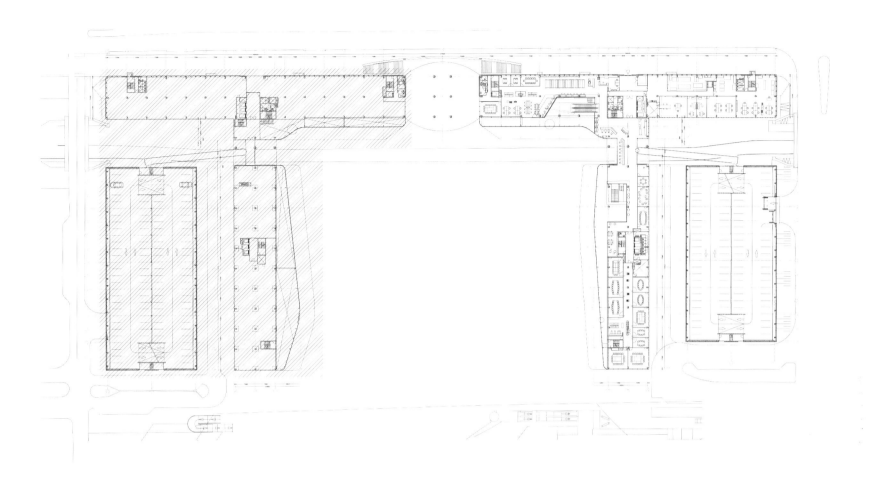

The 27,000 m² is spread over 6 floors. To create diversity and orientation in the long building the NUON Energy Sources were taken as a starting point in the interior design. These 6 sources were each translated into their own section design through the assignment of colour, material, photo prints and glass foils. This way, each of the building's "neighbourhoods" has their own specific identity, and a clear location within "nieuwAmsterdam".

With its renovation, the building has been certified "BREEAM Very Good". To achieve this sustainability class, points were scored both on the exterior and the interior design. The latter is remarkable because BREEAM has not yet been adapted to new workplace concepts, which means that the interior design has to meet high demands to be able to achieve this score.

The architectural renovation has been realized by the Architecten Cie, and led by Bramir Medic.

这个阿姆斯特丹东南部的项目通过可持续地再利用27 000m²的空间，重建了一种"新工作方式"。

这栋"nieuwAmsterdam"建筑建于25年前，最近转型成现代混合使用建筑，包括宾馆、餐饮业和NUON电力公司的27 000m²办公空间。

2012年9月，HEYLIGERS d+p建筑设计机构接受了为NUON设计室内空间的合同，在短短的15个月后，新办公室便投入了使用。新总部适应了"新工作方式"，移动办公：完全开放、极其灵活的办公环境。

室内设计了开放办公楼层，设置了多样灵活的办公空间，此外还穿插了会议室、电话间、单人工作区、食品室、"行政套房"和项目策划室。建筑为使用者提供了多样化的设施，包括接待大厅（带有水吧）、餐厅（可容纳650人）、图书馆、浓咖啡吧、服务中心、会议中心以及空中酒廊。

项目设计包括6层，总面积为27 000m²。为了让这栋长条形建筑内的空间更加多样化，导向性更加明确，NUON能源公司便成了室内设计的出发点。通过使用不同的色彩、材料、图案和玻璃，6种能源都被充分诠释到了各自的分区中，从而让建

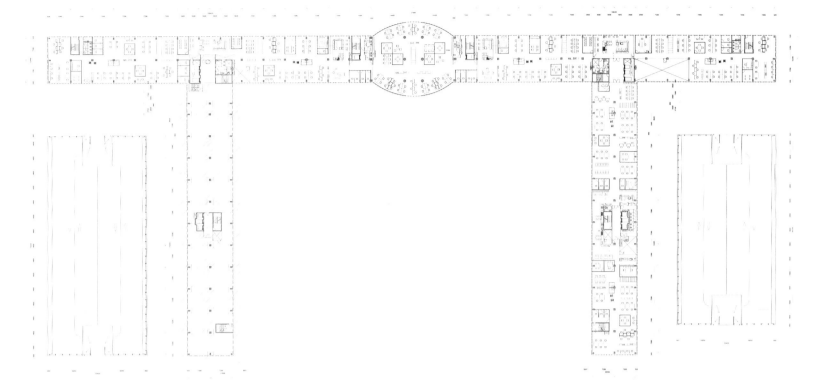

筑的每个"邻域"具有自身的特点,在"nieuwAmsterdam"建筑中的位置也十分明确。

建筑经过翻新之后,被评为三星级英国绿色建筑(BREEAM Very Good)。还对建筑的内外设计进行了评估,来确定建筑的可持续发展类别。后者更加有说服力,因为当时英国绿色建筑评估体系(BREEAM)还未应用到新办公空间理念上来,这就意味着室内设计必须满足很高的要求,才能获得这种评价。

建筑翻新项目是在 Bramir Medic 的指导下,由 Architecten Cie 完成的。

MANUFACTURING

Unilever Algida Ice Cream Factory

联合利华 Algida 冰激凌工厂

Design Agency: Studio 13

Designer: Deniz Yetkin

Client: Unilever

Location: Konya, Turkey

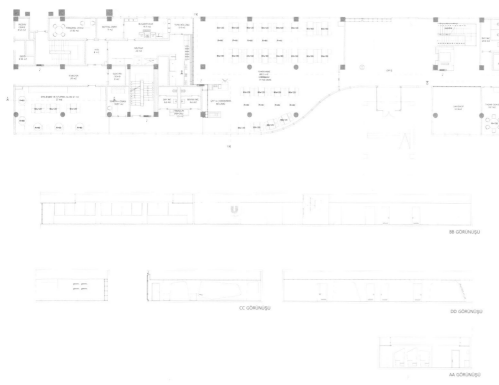

One of the best examples among Studio 13's many projects is Unilever Algida factory's offices and social areas design. The factory is located in Konya, Turkey and being a candidate Unilever's globally first factory with environment friendly LEED certificate. By designing all the offices and social areas, Studio 13 has made a very efficient design for employees with a vision of environment friendly green office.

With Unilever's sustainability vision, Algida Konya factory is currently using the following methods;

rain water harvesting and re-use, natural lighting methods in the production and storage areas, heat recovery in the productions areas and auxiliary plants, plant location optimization to reduce the carbon footprint, "zero waste" management, heat production from the waste treatment plant, onsite waste separation system, 100% recycled energy usage at process and auxiliary plants, energy monitoring and analysis system in auxiliary plants and production areas, using high efficiency motors and equipment throughout the factory. The interior design is completed based on this context by pay attention to these methods.

联合利华 Algida 工厂作为 Studio 13 众多的出色项目之一，位于土耳其科尼亚，是联合利华在全球的首家环境友好型 LEED 认证候选工厂。

Studio 13 创造了一个环境友好型绿色办公空间，不仅便于员工高效工作，也实现了节能和节水的最大化。

Studio 13 在设计科尼亚 Algida 工厂时，紧跟联合利华的可持续发展愿景，遵循了下列方法和原则：雨水收集和二次使用；生产和储存空间自然光照明法则；生产区和辅助厂房热量回收法；车间位置最优化以减少碳排放；"零浪费"管理法；垃圾处理厂产热；垃圾就地分类原则；加工和辅助车间能源全回收利用原则；辅助车间和生产区能源管理和分析系统；全厂采用高效发动机和机器。在此背景下，室内设计也从这些方法出发。

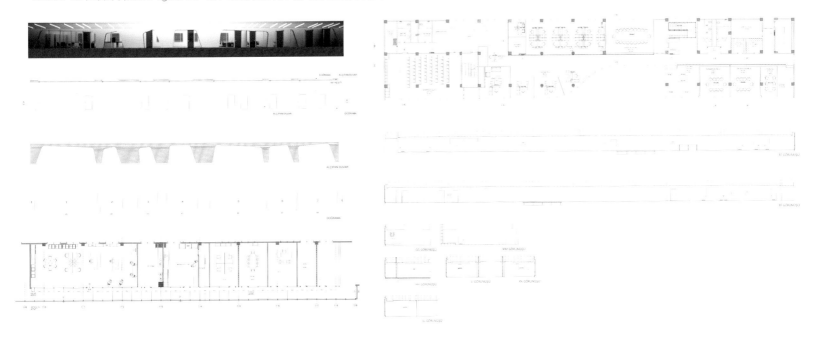

MEDIA/PUBLISHING

BBC Worldwide Americas

英国广播公司美洲总部

Architects: Perkins Eastman

Interior Design: Perkins Eastman

Client: BBC Worldwide Americas, Inc.

Location: New York, USA

Area: 3,716 m²

Drawing: Courtesy Perkins Eastman

Photography: Chris Cooper

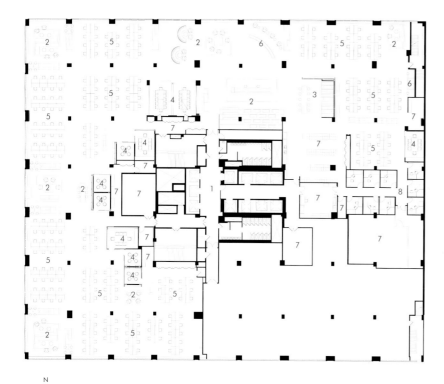

BBC WORLDWIDE AMERICAS floor plan

1 Elevator Lobby
2 Open Collaboration
3 Dining
4 Conference
5 Open Plan
6 Video Lounge
7 Support
8 Edit Suite

Designed by international design and architecture firm Perkins Eastman, the headquarters for the British Broadcasting Corporation (BBC) Worldwide Americas is a 3,716 m² loft-like space located in the heart of New York City's Times Square.

With the famed British media company's North America division quickly expanding its workforce, BBC sought Perkins Eastman's interior design and space planning services to consolidate their offices from 3 floors to 1 and to expand by nearly triple. A uniquely creative British presence in the middle of Manhattan provided ample sources of inspiration for the unconventional and collaborative environment. Marrying influences from the UK and the US resulted in a main reception area that incorporates bleacher-style seating with wing-back chairs, a 20' long Chesterfield sofa, and a curved LED video wall to highlight the vast array of BBC programming as guests enter the space.

Each of the 260 employees was assigned a seat along a benching system desk, or a "team table". Although the main headquarters in London had first adopted a completely open office policy, the US transition was surprisingly seamless despite some anticipated cultural adjustments. Senior staffs who were previously in private offices found that creativity and communication increased dramatically after adopting the benching system, thereby leveling the playing field. There are no dividers between or across users, which promotes communication, especially among the creative departments and ad sales. Wireless headsets allow employees to start a phone call from their desks and walk to a nearby meeting space without a break in the conversation. The surrounding open meeting

spaces were designed to be welcoming and flexible, breaking from a corporate aesthetic and embracing a more residential or hospitality aesthetic. The variety of collaborative settings range from living room like lounges and banquettes to bar-height tables that encourage quick interactions between passersby and those already seated at a barstool.

Closed meeting rooms incorporate a full complement of video conferencing and presentation tools alongside wall graphics that correspond to specific BBC programming or business lines. At times, quick interviews are shot with some of these spaces as the backdrop.

The reception area is used for town hall meetings with video conferencing built-in by placing cameras on columns and using the LED video wall as the remote video. Events and parties are occasionally hosted in the expansive reception area. On those occasions, the operable glass wall behind the reception desk is closed and the east side of the board room is opened to add a more continuous public space.

Unique and varied color, pattern and texture for furniture and light fixtures help identify different quadrants on the floor as the perimeter was kept free of any walls that would block views or natural light. Daylight dimmers help to control the lighting levels as the sun moves around the building. Paper use was reduced by roughly 20% as storage was reconsidered and printing rooms were centralized. The amount of carpet and ceiling is minimized to not only expose the waffle slab and newly polished concrete floors, but to reduce the amount of material that would need to be recycled. The decorative resin ceiling in the reception area and the colorful resin doors for conference rooms have 40% recycled pre-consumer content. All millwork utilizes FSC certified woods.

英国广播公司美洲总部是由国际设计和建筑公司Perkins Eastman设计的。这个面积约为3 716 m²的阁楼式空间，坐落在纽约时代广场中心。

随着著名英国媒体公司的北美分部迅速扩大其工作场所，Perkins Eastman设计公司被邀请来做室内设计和空间规划，并将3层楼合并成1层，使空间扩大3倍。这个不同寻常的合作环境的设计不断从"曼哈顿中心的创意独特英伦外观"的设计想法中汲取灵感。主接待区融合了英式风格和美式风格，并摆放着看台式座位和后靠式座位。

长长的办公桌可以容纳260位职员，也可称作"团队桌"。伦敦总部率先采用了完全开放式的办公政策，而美国办公室的转变也非常完美，除了部分事先做的文化方面的调整。高级职员原先是在私人办公室中工作，而现在也在长办公桌上工作，消除了等级差异，他们也发现自己的创造力和互动交流都大大增加了。员工的工作台之间不再有隔板，从而增强了员工之间的互动交流，尤其是创意部门和广告销售部门。职员都使用蓝牙耳机听电话，当需要离开座位到附近的会议区接听电话时，也无需打断谈话。四周的开放会议空间不仅灵活，也很吸引人，其设计打破了企业式风格，是一种温馨的家居风格。各种为合作设计的设施包括客厅式休息区、宴席长凳利与吧台同高的桌子，能让路过的人与坐在酒吧高脚凳上的人们很快地进入交流状态。

封闭式会议室内装有完整的视频会议设施和展示设施，而墙壁上的图案则反映了具体的公司项目和业务。短期会面有时会以这些空间为背景进行拍照留念。

接待区内配有视频会议设施，例如，柱子上的摄像头，以及远程视频会议LED视频墙，可以举行全体会议，各种大小活动和聚会也可以在这个宽敞的接待区中举行。在这些场合下，拉上接待台后的玻璃墙，打开东侧的董事会会议室，可以形成更大的公共空间。

由于空间内未设任何可能挡住视线和自然光的隔墙，所以采用了色彩、图案和质感各不一样的家具和灯具来区分不同的空间。日光调节器有助于控制空间的采光量。通过对纸张存放区进行调整，并把打印室集中在一起，纸张使用量也大约减少了20%。楼板和重新抛光的水泥地面都被裸露出来，使地毯和天花板的使用量也减到了最小，同时，需要回收的材料的数量也相应地减少了。接待区的装饰性树脂墙和会议室中彩色的树脂门重新采用了前废料的40%。除此之外，所有的木质设施采用的都是经过森林认证（FSC）的木材。

MEDIA/PUBLISHING

MTV Networks Headquarters

音乐电视网总部

Design Agency: dan pearlman Markenarchitektur GmbH

Designer: Marcus Fischer | Founder & Associate Creative Director Brand Experience & Exhibition

Client: MTV Networks

Location: Berlin, Germany

Area: Ground floor: 280 m²

First floor: 435 m²

Photography: diephotodesigner.de

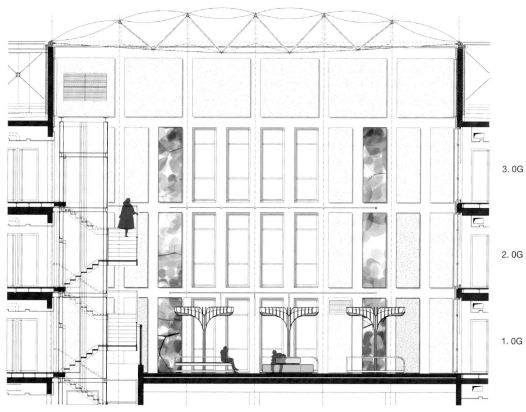

Working and taking breaks in style / work-life-balance at MTV Networks in Berlin.

Interior-design and re-organization of the reception, meeting and break areas of the new Headquarters of MTV Networks on the spree.

MTV Networks' personnel can rejoice: with the redesigned and newly organized reception, meeting and break room areas at MTV Headquarters, a long-held wish of employees becomes a reality and a once gloomy office complex on the spree can finally function as the head offices.

dan pearlman was asked to perform their magic and created an "axis of inspiration" by opening up the entire centre area on the first floor from the northern to the southern facade. Today, the spacious atrium serves as a "Brand Garden" with seating for taking breaks or informal meetings, which can accommodate up to 250 employees.

The interior design and materials relate to the identity of the MTV Networks (Germany) logo with its characteristic font and the colours dark brown, white, and yellow. They are referenced and translated

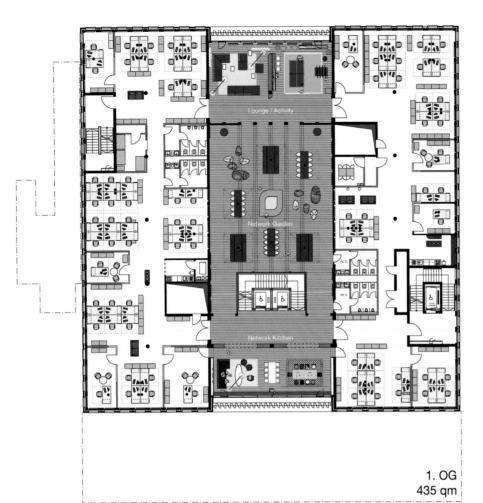

1. OG
435 qm

creatively in the design of the reception area, the lobby, and the café as well as in the atrium and the kitchenettes.

Greater freedom of movement and an equal variety of unique as well as private retreat zones contribute to a better work-life balance at the workplace. Whether eating lunch in the network kitchen, in the blue lounge or in the brand garden under a stylized canopy of leaves, there are ample choices for taking time out during the workday. Fans of table tennis or table football can of course be found in the sport lounge. The abstract tree parasols and several different seating isles refer to the self-contained "corporate" design and use of materials in the entrance area. The overall picture of the "garden" is rounded up by the interaction of the large airspace-leaves installation and the "vertical green" spaces as well as small details, such as felt stones as open seating isles.

For the realization of the project dan pearlman mainly worked with the two executive companies "artis möbel objekte raumkonzepte gmbh" and "KRAUSS Baugesellschaft mbH Berlin". The "KRAUSS Baugesellschaft mbH Berlin" (KRAUSS construction company) took the entire space-creating reconstructions of floors, walls and ceilings, from the color design to the implementation of the light. The "möbel objekte raumkonzepte gmbh" (furniture objects spatial-concepts company) realized the concept of the new work-life-balance at MTV with the whole interior decoration by stylish furniture and umbrellas. In addition, the entire team was supported by the Lindemann projekt.net GmbH & Co KG in researching and supplying the right furniture.

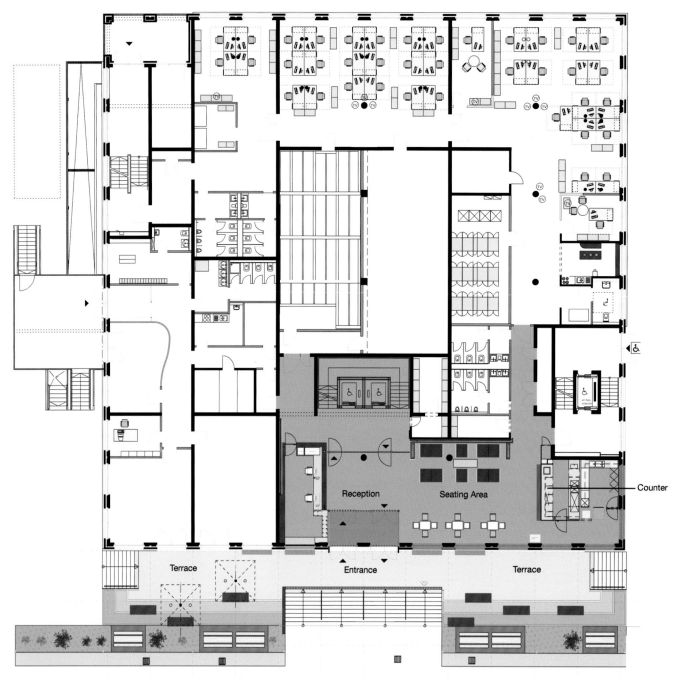

Ground Floor
280 sqm

有格调地工作；有格调地憩息，音乐电视网（MTV Networks）总部实现了生活与工作的平衡。

施普雷河岸的新 MTV 总部设计项目涉及室内设计以及接待空间、会议空间和休息空间的重新布局。

经过对 MTV 总部内接待区、会议区和休息区的重新设计和重新布局，职员们期盼已久的事情成为了现实，施普雷河岸曾经的昏暗办公楼群终于成为了总部，这令 MTV 的职员无比欣喜。

dan pearlman 建筑设计机构接手了总部的设计，他们大展身手，将一楼的从北至南的整个中央区域敞开来，创造了"灵感之轴"。如今，宽阔的中庭成为了公司的"品牌花园"，还为职工休息、非正式会议提供了座椅，可容纳多达 250 位职员。

室内设计风格和材料的选择都与 MTV（德国）的标志特征相符，例如，字形特征和暗棕色、白色和黄色的用色方案。接待区、大厅、咖啡区、中庭和厨房的设计很有创意，很好地诠释了公司的形象。

新办公区拥有更多的移动空间，多样的私人休息区，具有同等的独特性，更好地平衡了生活与工作。不管是在公司厨房，还是在蓝色休息区，或者是在品牌花园中的独特树叶华盖下，工作之余可选择的休憩地点总是数不胜数。喜爱乒乓球和桌上足球的人们当然会聚集在运动休闲室。抽象的树冠和各种单独摆放的座椅，以及门厅处材料的选用，都反映出独一无二的企业设计。悬浮于空中的树叶大灯、"垂直绿色"空间，以及诸如充当独立座位的石头之类的细节，点缀了空间，让这个"花园"更加完美。

为了更好地完成项目，dan pearlman 设计机构主要与两家项目执行公司进行了合作："artis möbel objekte raumkonzepte gmbh"和"KRAUSS Baugesellschaft mbH Berlin"。前者是一家空间构思和家具陈设公司，它用时尚的家具和伞具装饰了空间，将 MTV 的生活与工作平衡的概念融入空间中。后者是一家建设公司，负责了整个空间中的地面、墙壁和天花板的重建，甚至还有用色方案和灯光应用。除此之外，设计团队还得到了 Lindemann projekt.net GmbH & Co KG 的支持，包括家具的定制和供应。

MEDIA/PUBLISHING

Wunderman/Bienalto Sydney

悉尼Wunderman和Bienalto公司办公室

Design Agency: The Bold Collective

Interior Designers: Monika Branagan, Ali McShane, Therese Lowton, Kristy Lee

Street Artists: Numskull & Beastman

Photography: Andrew Worssam

Area: 930 m²

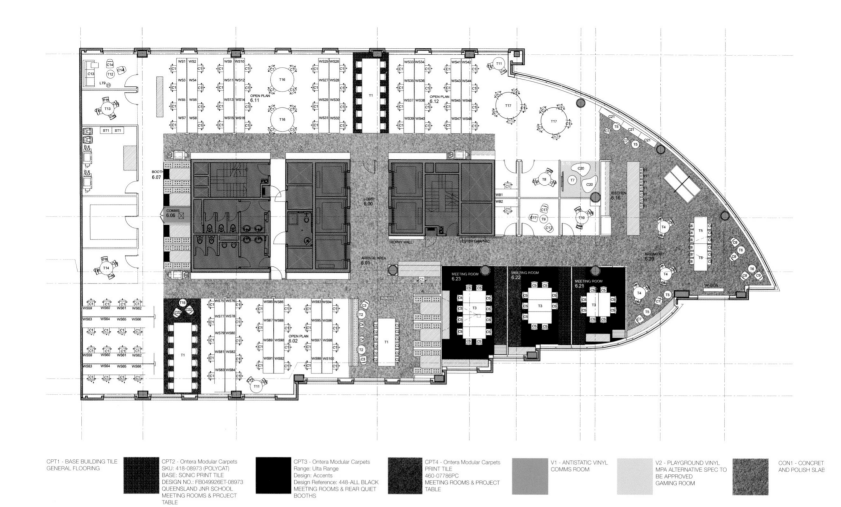

Overview:

Media Agencies, Wunderman and Bienalto were to move from 2 floors in Darlinghurst into 1 floor at 35 Clarence Street, where they would exist in the same building as some of the sister companies under the WPP umbrella.

They were after a bright and inspiring new work place that would excite staff and clients alike. They were keen to explore an Activity Based Work setting that responded to changing workplace conditions and that would enhance workplace culture. As an Activity Based Working project, we had to allow for a variety of work settings including seamless partition-less bench style workstations, high bench style project tables, enclosed booths, small quiet rooms, meeting rooms as well as an active work-friendly reception area and large communal breakout that would be able to host large meetings.

The design response was bright and playful. Our aim was to break away from tradition, so we experimented with materiality and colour. Drawing on the colour palette from the company logo, local graffiti artists were engaged to paint the arrival area to an immediate sense of the media agency's brand for any visitors.

Brief:

Wunderman and Bienalto were forward thinking and keen for us to pursue an Activity Based Workplace. This meant that there would be no assigned seats and the staff would need to be provided with a variety of functional settings to work in. The floor would be highly populated so it would be important to create environments that allowed staff to escape for a private phone call, collaborate with other staff, meet with their clients, and to relax on their lunch break.

In addition to this, the new workspace was to enhance the company culture and bring the 2 companies who had previously occupied separate floors, together. As a part of this brief, we were to incorporate a number of existing elements iconic to the clients. These included a large purple "W", rainbow coloured stairs adorned with hand written messages, stenciled table tops, a large perforated portrait of the company founder and a series of purple gnomes to name a few. The client also wanted to be projected as a leading professional agency.

Innovation:

It was important that we moved away from tradition, explored spaces that encouraged activity based working, and focused creating both an exciting place to work and a functional office that would transcend time and welcome external stakeholders.

We invited local street artists to paint the lobby arrival area, equipping them with a palette of corporate colours. The reception desk sits proud of an open work setting consisting of banquette seating, a collaborative work table, and enclosed booths so that staff and visitors alike can utilise. Post-occupancy we have found this to be one of the most successful spaces and it works well to create a "buzz" in the office. A corridor leads you away from reception, along a series of quiet and meeting rooms as well as the operable boardroom, through to a large communal breakout with views toward the Sydney Harbour Bridge. This space hosts large meetings and is littered with loose furniture, a ping pong table and foosball. The large kitchen services the entire office and has the best view in the fit-out.

Challenge:

The project timeframe was very tight. The builders had 6 weeks to produce the fit out which had large amounts of bespoke joinery that would accommodate the activity based work settings. On top of this, we had to select furniture and finishes that could be delivered under such a tight deadline. It was a great challenge not to compromise the design given the restrictions on the project and we worked hard with the builders to resolve details on site and push suppliers for fast lead times.

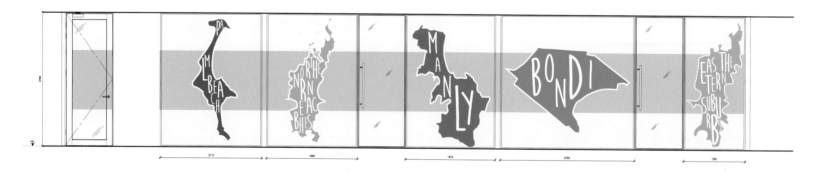

概观：

Wunderman 和 Bienalto 两家媒体公司打算从达令赫斯特的 2 层楼办公室，搬至克拉伦斯街 35 号单层办公室中，与 WPP 集团旗下的姊妹公司在同一栋楼中办公。

这两家公司想要一个能启发职员和客户灵感的明亮空间。他们热衷于探索，能够适应变化的工作环境和增强工作场所文化活动的办公空间。一个活动办公空间应该设有多样的工作设施，例如，无隔墙长办公台、会议室和充满活力的办公的接待区，甚至还包括能够举办大型会议的公共休息区。

针对这些要求，项目采用了明亮和俏皮的设计风格。设计师尝试了不同的材料和色彩，旨在突破传统。还从公司的商标中获取用色灵感，并请当地的涂鸦大师在入口区绘画，让访客直接感受到媒体公司的品牌力量。

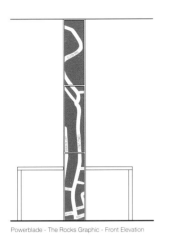
Powerblade - The Rocks Graphic - Front Elevation

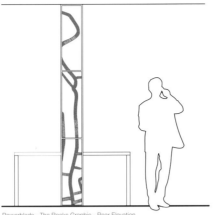
Powerblade - The Rocks Graphic - Rear Elevation

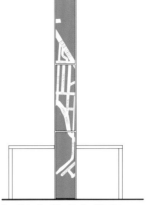
Powerblade - Bondi Graphic - Front Elevation

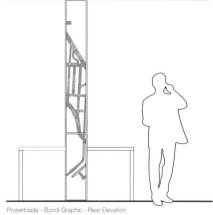
Powerblade - Bondi Graphic - Rear Elevation

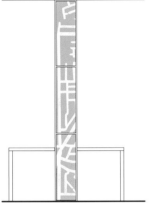
Powerblade - Kings Cross Graphic - Front Elevation

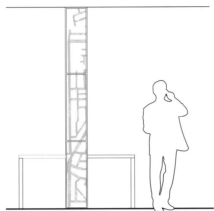
Powerblade - Kings Cross Graphic - Rear Elevation

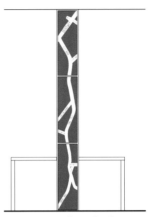
Powerblade - Newtown Graphic - Front Elevation

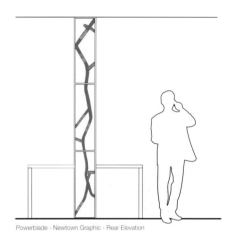
Powerblade - Newtown Graphic - Rear Elevation

This would ensure that our client achieved their high density workplace and parent-company requirements with the level of design features that would make the space come to life and inspire staff every day.

Sustainability:

As the project was located within a typical office building, we got in early and collaborated with building management to ensure the "make good" was in keeping with the design intent. By doing this we minimised material waste and labour resources. Carpet was laid only where it was needed and ceiling tiles were not applied to the entire floor as our design intent was to remove a large portion of both. We detailed a considerable amount of bespoke joinery to be made by local tradesmen, rather than specifying off the shelf products, and we looked to local suppliers for furniture specifications where possible. A difficult task given the timeframe we were working with.

The majority of the built environment was kept to the core, which allowed the open workspace and breakout areas to utilise the natural light from the glazed building facade. We kept the finishes light and airy and included plant life to bring biofilic quality to the space for the wellbeing of staff. Planting in interior environments eliminated VOC's, vastly improving the surrounding air quality. This brings direct benefits to the occupants of the space.

Awards:

Best Commercial Interior at the Sydney Design Awards 2014

简述：

Wunderman 和 Bienalto 向我们表达了其对"活力办公空间"的想法和期望。这就意味着职员不再有固定座位，取而代之以各种功能设施，以满足职员的工作需求。然而，在这样的模式之下，楼层上人员流动会很大，因此为职员接听私人电话，或者与其他职员协作，会见客户，甚至午休等创造相应环境也极为重要。

除此以外，新办公空间应当巩固企业文化，将之前各占一个楼层的公司凝聚在一起。我们还需将客户原有的一些标志性元素加入到设计中，包括紫色"W"字母；被手写留言装饰着的彩色楼梯；印花桌面；公司创始人大穿孔肖像以及一系列的紫色格言，等等。同时，客户计划发展成为领先的专业机构。

创意：

本项目的重点是打破传统，探索充满活力的办公空间，专注于创造令人振奋的、功能性的工作空间。从而使空间能经久不衰，吸引外部人员。

受邀的当地涂鸦艺术家将大厅入口区域涂画成彩色的图案。接待台在开放的工作设施中十分突出，这些设施包括直排座位、协作桌和封闭式空间，职员和访客都可以使用。搬进新公司后，我们发现这是最成功的办公空间之一，它能促进人员之间的交流。从接待区的过道可以直接通往宽敞的公共休息区，在这里可以远眺悉尼港桥，而走廊两侧则是较安静的房间、会议室和可拆除会议室。大部门的会议室都设在这里，各种家具零散地摆放着，包括乒乓球台和桌上足球等。大厨房服务于整个办公室，其装配十分壮观。

挑战：

项目安排时间紧促，施工人员仅有6周的时间来进行生产装配，其中包括大量与活力办公布置相配的定制细木工家具。更重要的是，设计师还需挑选能快速交货的家具和油漆。在诸多项目条件限制下一丝不苟地设计是一种挑战。设计师还与施工人员现场解决细节问题，催促供应商交货。因此，一个符合母公司要求的密集型办公空间才得以如期完成。空间的设计具有特色，贴近生活，激励着职员。

可持续性：

由于项目位于典型的办公楼中，设计师提前与楼房管理部门沟通，确保设计过程顺利进行。我们将物力和人力浪费降到了最小。地毯也仅仅铺在有需要的地方，整个屋顶也并未完全用吊顶板覆盖起来，恰好与设计计划相符。我们从当地商人手中采购了大量的定制木工制品，而非采用货架上现成的产品，还跟进家具的制作规格。最终，我们完成了这个时间紧促的艰难项目。

大部分的建筑环境都设在中央。因此，开放工作空间和休息区能充分利用从玻璃墙照射进来的自然光。室内装饰明亮、通风，植物给空间带来了生机与活力，增加了职员的幸福感。这些植物还有利于吸收有机挥发物，提高周围空气的质量，直接让空间使用者受益。

获得奖项：
2014 年悉尼设计大赛最佳商业室内设计

MISCELLANEOUS

'T PARK

T公园办公空间

Architects: Pieter van der Pot (CUBE Architecten), Marloes van Heteren (CUBE Architecten)

Client: CITY OF AMSTERDAM (PMB)

Area: 530 m²

Photography: Yvonne Lukkenaar

INSIDE/OUTSIDE OFFICE DESIGN V

In August 2013 the City of Amsterdam gave the go-ahead for the redevelopment of an old store room under one of their offices on Jodenbreestraat in Amsterdam into a multifunctional plaza. This plan is a part of the new flexible housing concept for their offices, which can be densified through realization of such plazas. CUBE is, after a closed competition, asked to develop this 530 m² plaza. According to the City of Amsterdam it had to become an attractive area with an extraordinary appearance.

The semi-public plaza can be used for working, collaborating, meeting, or eating and drinking the organic coffee, juices and sandwiches from the bar. There is also the possibility for holding presentations and meetings. The more functional parts of the program are efficiently packaged in simple volumes in such a way that they form an open space for the plaza. The entrance area of the huge office building was redeveloped with a new reception desk and added to this new plaza. Although the fire separation in the building needed to be changed for this, the open connection between the entrance and the plaza was really important to make the new addition really part of the whole building.

By this we have added a space to the building with the peace and playfulness of a green park. Large different plant cages hang like a canopy between the real birch trunks. Along the large glass doors planters are hung that form a natural filter. On the walls a print of the shadows of the leaves was used, and the floor has a natural outside look: with its gray / brown tones. The consistent use of the colors white, brown and green brings certain calmness between the otherwise fairly random placed elements. This seemingly random placement of trees, plant cages and furniture is in reality a precise positioning relative to the projector and the walking paths. The volumes containing extra spaces that are located as wooden sheds between the trees, have wooden slats of unequal thicknesses.

The real plants in the hanging cages, which from the beginning were an essential part of the plan and ensure good air quality and a green experience, were the biggest challenge in the design process. Initially we looked at an automatic watering system, but in the end we opted to hang them on 10 electric hoists that are normally just in theaters. Every 4 weeks they will be lowered to take care of the hydroponic plants.

Sustainability was one of the key principles and reuse is a part of it. We could reuse parts of the installations, and also for the furniture we went looking for used elements in the storage depots of the municipality. The pendant lights above the counter and some of the tables and chairs all come from here.

2013年8月，阿姆斯特丹市批准重新开发一家旧商店使之成为多功能广场，该店位于阿姆斯特丹犹太大街上，恰好位于市政办公处楼下。办公室采用了新的灵活办公空间理念，并可通过改造商店来丰富空间。CUBE建筑公司通过了激烈的竞争，应邀开发这个530m²的空间。根据阿姆斯特丹市政的要求，这个广场应当具有与众不同的外观，能够吸引人们的目光。

这个半开放的广场可以用来工作、合作、召开会议或作为酒吧、用餐、休闲场所。同时还可以用来举办展览和活动。照此，更多的功能空间也都有效地设在了这个简单的空间中，从而把广场转变成了开放空间。大办公楼的入口处设有接待台，因此建筑的防火隔墙也需做相应的改变，但入口与内部间的连通性十分重要，因为只有这样，新增的部分才能成为大楼一部分。

我们为这个建筑增加了一个安谧而充满活力的绿色花园。各种不同的大植物笼子吊在白桦树之间，就像是桦树树干上方的树冠。沿着大玻璃门，悬挂着许多盆栽，形成一道天然屏障。墙上贴有树叶阴影图案墙纸，地板具有天然灰色或棕色外观。随意摆放的物品仍然采用了白色、棕色和绿色色调，营造出静谧的空间氛围。这些看似摆放随意的树木、植物笼和家具，实则定位精准，与投影仪和走道联系密切。树木间的木棚是由薄厚不一的木条做成，这些额外空间对整体空间起到了补充作用。

悬挂的笼子里的植物一直是设计的重要成分，它有利于保证空气质量，给人以绿色空间的体验，同样它也是设计过程中的最大挑战。最初我们考虑采用自动洒水系统，但最终选择把它们悬挂在剧院中常见的电动起重机上。每隔4周起重机就会被放低，以便给这些植物浇水。

项目的核心设计理念是可持续性。当然，再利用原则也是不可忽视的。我们重新利用旧设施，并从市政仓库中寻找了些家具，例如，柜台上的吊灯、部分桌椅和其他家具。

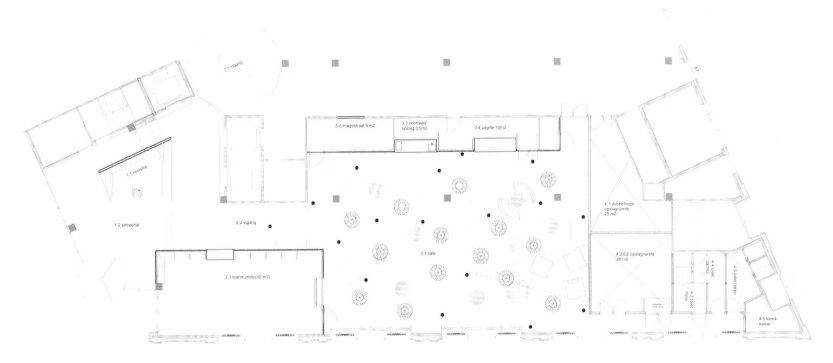

Plattegrond

Integral Iluminación Comercial Building

Integral Iluminación商业大楼

Design Agency: Jannina Cabal & Arquitectos

Lead Architects: Arq. Jannina Cabal

Design Team: Arq. Alejandra Lopez, Arq. Marco Sosa, Arq. Katty Cuenca

Location: Samborondon, Ecuador

Construction Area: 12,701 m²

Photography: Juan Alberto Andrade, Sebastian Crespo

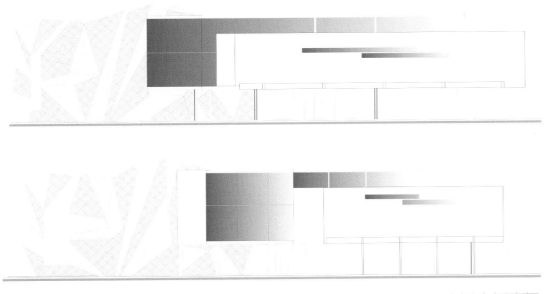

General Parameters
Oriented from north to south, Integral Iluminación Commercial Building, developed in 2 diamond-shaped plots with a total building area of 1,180 m².

One of the requirements of the project was to keep the exterior structure of the existing house (columns, walls and pitched roof) in one of the plots while a new structure proposed for the expansion of the building on the second plot.

Considering that Integral Iluminacion sells top of the line lighting and automatization systems, it was extremely important to conceive a cutting edge architectural design that would become an icon for the company.

Architectural Program
The architectural program of the project developed as a 3-storey building. The first level contains the

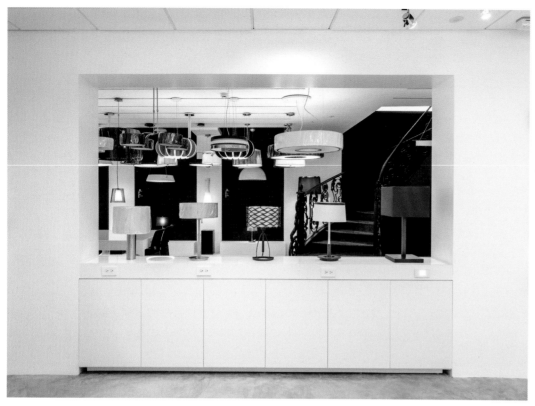

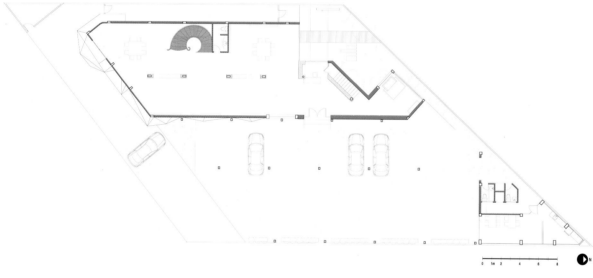

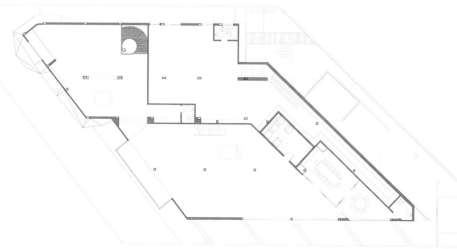

常规参数

Integral Iluminación 商业大楼坐落在 2 块由北至南铺开的宝石状场地中,总建筑面积约为 1 180 m²。

按照项目的要求,一块场地中的原建筑的外部结构(柱子、墙壁和斜屋顶)被保留下来,而第二块场地中的建筑被扩大,形成了新的结构。

鉴于 Integral Iluminacion 是专门提供高端灯具和自动化系统的公司,因此构思一个棱角鲜明、能够代表公司形象的建筑结构极为重要。

建筑规划

建筑分为三层,第一层主要设有停车区、小前厅、室内花园、展示空间和职员餐厅。第二层是 2 倍高展厅,展厅内设有销售台、影视间、大小会议室和储藏室。而第三层的整个空间都被用作了管理、行政和灯具设计办公区。

布局方案

项目的特色就是两种元素的融合。左侧的原结构被分隔成不规则三角形的弹性几何形织物包裹起来。这种外部结构是由钢管和铝管

parking area, a small lobby overlooking an interior garden, exhibition space and employee dining room. The second level has a double-storey exhibition space with sales desks, a cinema, meeting and conference rooms, as well as storage space. The entire third level was developed for the management, administration and lighting design offices.

Compositional Scheme
The project is clearly defined by the merging of 2 different elements. On the left side, the existing structure wrapped up in a tensile fabric with a geometric - based pattern decomposed into irregular triangles. The facade structure made of a mixed - metal structure of steel and aluminum tubes wrapped in a medium gray tensile fabric developed specifically for outdoor use only. While on the right side a clean-cut double- story glass prism is raised up on an exposed metal structure creating a covered garage underneath.

The perception of the building varies completely from day-to-night. During the day the tensile fabric facade is a visual element which does not compete with the glass cube of the expansion of the building, creating a homogeneous structure altogether. At night, colored lighting brings the building to life highlighting the metal structure and accentuating the three dimensional effect of the panels in a game of lights and shadows. The building is therefore transformed into a sculpture. This mesh fabric, permeable to light, lit with RGB LED reflectors allowing the facade to change colors. The reflectors are connected to an automatization system that controls the building to change through the night.

Interior
The concrete floors, white walls and white furniture reflect interior spaces conceived seeking greater simplicity. Even though this is a retail space with few windows to the street, the interior opens up to naturally lighted spaces by an interior patio and a fountain, focal elements on the ground floor.

These simple fresh and generous areas highlight the vertical circulation of the building, the oval staircase, the only architectural element maintained from the original house. The granite floors and black walls contrast with the neutral color scheme of the interior spaces.

组合而成的金属结构，结构上再用户外专用灰色拉伸布料包裹起来。而右侧是干净利落的两倍高玻璃棱柱状建筑，棱柱结构下方是暴露的金属结构，充当了遮阳车库。

无论是白天还是黑夜，建筑都会给人以不同的感觉。白天张拉布结构冲击着视觉，但是却不会夺去立方体玻璃建筑的光彩，两者互相呼应，十分和谐。夜晚，彩色灯光衬托着建筑的金属结构，烘托出外表的三维效果，赋予了建筑活力与生机。此时，建筑就像一座雕塑，庄严美观。灯光透过网眼布，照射到RGB LED反光镜上，这些反光镜在自动化系统的控制下，使建筑表面色彩发生改变。

室内设计
混凝土地面、白色墙壁和白色家具的搭配使室内空间显得十分简洁。尽管商店区朝向街道的那侧所设的窗户很少，但是室内露台和喷泉的设计给室内空间带来了充足的自然光，让室内空间显得十分开敞。

这些简洁、清新、宽敞的空间使建筑的上下流通自然。原建筑唯一保留下来的结构是椭圆形楼梯。其花岗岩地板和黑色墙壁与室内空间的浅色调形成鲜明的对比。

EMKE Office Building

EMKE 办公大楼

Refurbishment Design: LAB5 architects

Client: AEW Europe

Location: Budapest, Hungary

Area: 9,100 m²

Photography: Bujnovszky Tamás

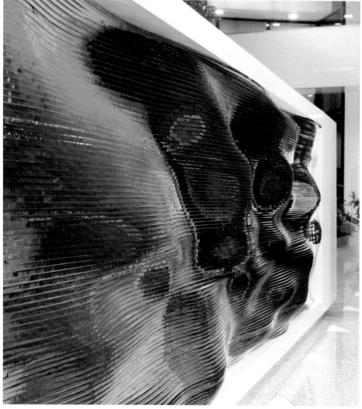

Location and Program
EMKE Office Building was one of the first of the new generation and Western-type public buildings in Budapest, when built in 1992. Its owner decided to refurbish it, modernizing the full MEP system, and providing a small new tenant area on the ground floor reachable directly from the street.

LAB5 architects proposed to relocate the main entrance and the access to upper floors. So visitors get a better impression when waiting for the elevator in the eminence of the glass-roofed atrium of large internal height. This way the new function could also be placed at the corner of the building, having an emphasised visual appearance from the square and the streets.

Interior Design
The only one and big change of the atmosphere is due to the generous application of new green elements. From one side, the interior becomes more of a transition between public and office zones, providing a semi-public space. From the other side, it is also a reference to nature; the building core becomes a green island in the city, to refresh the office workers and their clients or visitors.

The concept was to create the biggest possible space for the public functions, so all areas opened together, many times not only in visual solutions, but also in accessibility.

As the majority of the interior was meant to be kept untouched, only some small interventions were applied for the rest of the design proposal. Generally the space is kept whitish in aspect of restored elements, and all added objects are individuals, slightly stepping back from total integration, but never stating a contrast.

Materials, Shapes, and Colors
Nature is applied at many parts of the interior.

The waiting area as individual free-standing furniture, looking like an object of up-folded sheets, includes green surfaces of plants to hide the possible new function in the back, and to elevate the lounge to become more familiar.

The reception desk includes a glass surface of curvy cut to invoke waves of a river that demonstrates this point of the building as constant and intense flow of people, of information, and of communication.

Many sides of the atrium are covered by vegetables, just as the slope around the edge of the atrium slab. The railing looks like an exterior element could appear in any fields of small garden of agriculture.

The basement restaurant is shaded by 3 trees, to have an impression for costumers of being in an outdoor garden for lunch. The trees are elevated so they don't occupy space on the basement level, and appear above the ground floor as well.

As the space is deep counted from the glass roof of the atrium, we applied as many glass as possible. Walls are covered by white glass everywhere; expect the wall of the elevator, where it becomes black. For better orientation anywhere in the house this exception marks the vertical communication of the building.

Because of the new layout, the elevators are accessible from the atrium. Their shaft's new barrier wall and the sides of the elevator itself are made of frame-less glass, even their load bearing structure.

The mezzanine level functions are hidden in "flying boxes" that have a glass composition cover of white, sheer, and mirror glasses. As the office building if iconic because of its rounded glass wall corner street facades, we applied this as rounded triangles in the interior where felt necessity of a pattern.

MEP
The majority of the mechanical system of the building was redesigned and updated to recent and possible future expectations. New air-machines with heat-exchangers, new economical boiler, and new chillers are installed on the roof, and provide more environmental-friendly functioning. The ventilation, the FCU's and ducts in the interior are upgraded too where possible. A new BMS and week current power system is built to control the mechanical, the fire safety, and the electrical system.

定位与规划
EMKE 办公大楼是布达佩斯首批新一代西式公共大楼之一，建于 1992 年。房主决定翻新大楼，更新整套 MEP 系统，并在一楼设立从街上可以直接进入的租售区。

LAB5 建筑公司提议调整大门和楼梯的位置，从而给在高大的玻璃顶中庭中等候电梯的人们留下深刻的印象。如此一来，新的功能区将会设在大楼的一角，以突出从广场和街上观看大厅时的视觉效果。

室内设计
大量的绿色新元素被应用到空间中来，大大改变了空间的氛围，这也是空间的唯一的不同之处。一方面，室内空间变成了介于公共空间和办公空间之间的半公共空间。另一方面，空间具有自然性，大楼中心成为了城市中的绿色岛屿，有助于职员、客户和访客消除疲劳，抖擞精神。

项目的理念是设计尽可能大的公用空间，因此，所有的区域都互相开放，不仅

视线毫无阻碍，而且可以互通。

建筑内大部分空间都维持原样，只是按照设计方案稍做了更改。空间整体保持了白色外观，而增加的物品都是独立的部分，仅处于次要地位，不会夺去整体空间的光彩。

材料、形状和色彩
室内许多地方都具有自然性。

等候区是一整套独立的家具，看起来就像折起来的板片。绿色墙面遮挡了后部的新功能区。而休息区的位置较高，使空间的氛围更为亲切。

曲面玻璃接待台让人联想到波浪，代表着建筑内不断流动的人群，不断传播的消息和频繁的交流活动。

中庭的许多地方都被绿植覆盖着，中庭地板边缘的斜坡就是如此。而室内的栏杆就像是随处可见的户外花园的篱笆。

地下室中的餐厅设有3棵树为其遮阴，给人以一种在露天花园中用餐的感觉。这3棵树的位置较高，因此不会占用地下室的空间，而且树还高过建筑物的第一层。

透过中庭的玻璃屋顶往下看，空间显得很深邃，因此空间大量采用了玻璃。连接着各层楼的电梯间的黑色墙壁则具有导向作用，除此以外，其他墙壁也都装上了玻璃。

空间经过重新布局，可以直接从中庭通到电梯间。电梯间的保护墙和电梯各侧的壁板，甚至是承重结构，都是由无框玻璃组成。

夹层中的功能区都隐藏在由白色、透明和镜面玻璃组成的"飞行方盒"中。建筑的弧形玻璃幕墙恰好朝向街道，使建筑成为了具有代表性的办公大楼。而室内空间恰好也像是有弧度的三角形空间。

工程设施
建筑内的大部分机械设备都被更新，以满足目前和未来的需求。自带热量交换器的新通风设备、新节能热水器和冷却器都设在屋顶，具有更多的环保功能。通风机、风机盘管和通风管道都升级到了最新版本。新电量管理系统和最新的电力系统也都被用来控制各种机械、防火安全设备和电气系统。

MISCELLANEOUS

Mi9 Sydney

Mi9 悉尼办公室

Design Agency: The Bold Collective

Interior Designers: Monika Branagan, Ali Mcshane, Kristy Lee, Therese Lowton

Street Artists: Beastman And Numskull

Photography: James Macree

We were engaged by Mi9 to refit their existing office space at Australia Square with the aim of increasing the profile of the company core values for both staff and visitors alike. Drawing on the company manifesto, we extracted the key core values of "Humble", "Brave", "Give a Shit" and "Smart" and collaborated with local street artists to reference them throughout the space. The client wanted us to explore the idea of old and new technologies which we translated in a number of different ways which predominantly looked to juxtaposition and contrast as design elements.

The reception was located on level 6 which was to retain lofty heritage ceilings originally planned by Harry Seidler. As part of the fitout, we gave the ceiling a facelift by re-lamping the lights to brighten the floor and highlight the white modernist ceiling. We then worked with the existing structural elements, giving them a post-modernist aesthetic and providing them with awnings to highlight the contrast between the base building and new fitout.

Being a round building, the space was a challenge to plan but we overcame this by working with the client to eliminate hierarchy and deliver an open plan environment. We managed to utilise much of the existing structure to provide a variety of meeting rooms throughout the floor. We also created a number of informal settings so that the workstations were broken up, creating a rich composition of areas. Level 7 was a similar formula to level 6 however there was no heritage ceiling. Instead, we referenced the floor below by revealing the building structure above the large communal breakout area by removing the ceiling. The use of juxtaposing materials and street graphics referencing the company manifesto continued on level 7 to maintain consistency across the floors.

Our design honoured the base building and its heritage features. It became an asset rather than hindrance to our design solution. This included re-lamping the heritage ceiling on level 6 to energy efficient globes and upgrading the bath rooms. By honouring the site and upgrading it where we could, we provided the opportunity for part of this fit out to live beyond the years of its current tenancy.

The new lighting made the space much brighter, as did the use of white paint to bounce light through the space. Partition-less bench style workstations and an open plan setting also aided the light on the floor. Additionally we provided staff with their own energy efficient task lighting as to make it a more comfortable and sustainable environment for them to work in. The bench style workstations will provide our client with the room to grow as more staff can fit along each length of bench.

Much of the built environment was exiting so we eliminated the need for unnecessary demolition and construction and reused as much of the finishes and furniture that we could that was still functional and in keeping with the design concept, which lent itself to re-use as contrast and juxtaposition would play a key role.

Our material selections were locally sourced where possible and we also made use of reclaimed materials, including timber sleepers at the reception. We collaborated with local street artists to create engaging graphics through the floors. This type of collaboration is important for economic sustainability within the local context of Sydney. The client pursued the street artists for further work as they wanted to strengthen their manifesto branding.

该项目是为 Mi9 翻修位于澳大利亚广场的原办公室，旨在向职员和访客等人展示公司形象，提升公司核心价值。从公司的训言中，我们提取了公司的核心价值，这些关键词包括"谦逊、勇敢、洒脱、机智"，除此之外，我们还与当地的艺术家合作，将这些价值应用到空间中。客户希望我们能探索关于旧科技和新科技的概念，我们用并置和对比等不同的方式诠释这个设计元素。

接待台设在第六层，意在保存哈里·西德尔设计的老式高天花板。因为这一层也是项目的一部分，因此我们给它稍做了改变。经过重新布局的灯具让空间更加明亮，烘托了白色的现代天花板。针对原来的结构上的元素，我们赋予了它后现代主义的美感。通过创造一个遮阳篷状的结构来衬托原建筑与新结构之间的对比效果。

圆形建筑的室内空间很难规划，然而，通过与客户合作，消除等级结构，一个布局开放的工作环境便形成了。另外，通过充分利用现有结构，我们还在楼层上设立了各种会议室。各种非正式空间

将工作区分隔开来,形成了一个多分区的区域。第七层与第六层基本相同,而不同之处在于这一层没有传统天花板。我们则按照楼下的布局,通过移除天花板来突出大公共休息空间上方的建筑结构。第七层中的并列材料和街头图画,与地板保持一致,体现了公司的核心价值。

该设计尊重了原建筑以及它的传统特征。它对设计策略来说是一笔财富而非障碍物,比如第六层天花板上灯具的重新布置,节能地球仪的安装,以及卫生间的升级。通过适当保留和改造空间,使办公室配置的寿命能够长过目前的租期。新灯具使空间更加明亮,白漆的使用能将灯光反射到整个空间。无分区工作台和开放式的布局也增加了空间的亮度。而且,我们还为职工提供了节能工作灯,让他们在工作时更加舒适,更加环保。这种长凳式工作台允许公司按照工作台的长度来相应增加职员人数。

在该项目中,大部分的原有的建设都被保留下来,减少了不必要的拆建,还尽量采用了与设计理念相吻合的有用的旧装饰和旧家具,从而令这种对比和并置的手法也能在其中发挥关键性作用。

我们尽量从当地挑选材料,充分利用回收材料,比如接待台采用的枕木就是回收材料。我们还与当地的街头艺术家合作,在空间创造动人的图案。在悉尼的地方背景下,这样一种合作方式对经济可持续性具有重要意义。客户为今后的工作,追求街头艺术,旨在增强品牌的影响力。

Uniform

Uniform 办公空间

Architects: Snook Architects

Client: Uniform

Location: Liverpool, United Kingdom

Gross Floor Area: 650 m²

Photography: Andy Haslam

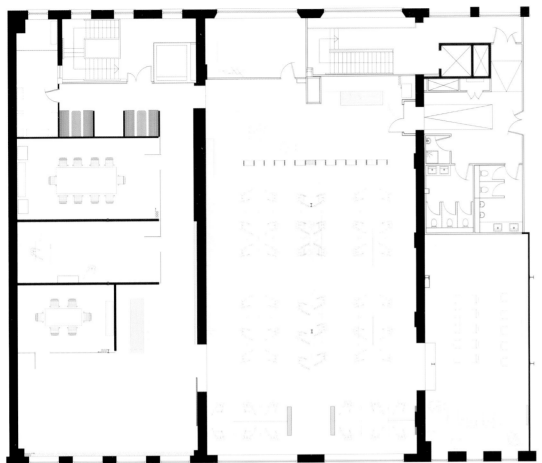

Uniform's work is eclectic in approach responding to individual clients' requirements in a surprising and often unexpected way. Their new office is a physical realization of this working method characterized by the ever changing media wall that acts as both a separation from the reception area to the studio space beyond. Rather than a complete physical barrier the wall offers tantalizing views to the work being done beyond and acts as a platform to present work to potential clients.

The overall palette of materials used in the project continues the inventive philosophy utilizing everyday utilitarian fabrics in unexpected situations.

Uniform 公司专注于用独特的方式来处理特殊客户的要求。公司的新办公室是这些工作策略的具体体现，千变万化的媒体墙既充当了接待区的隔墙，又分隔了工作区，成为了空间的亮点。墙壁不单只是简单的实体屏障，还展示了楼上的工作场景，激发了人们的好奇心。此外，它还是为潜在客户展示工作的平台。

项目采用的材料遵循了一个创意理念：把日常的实用材料应用到意想不到的地方。

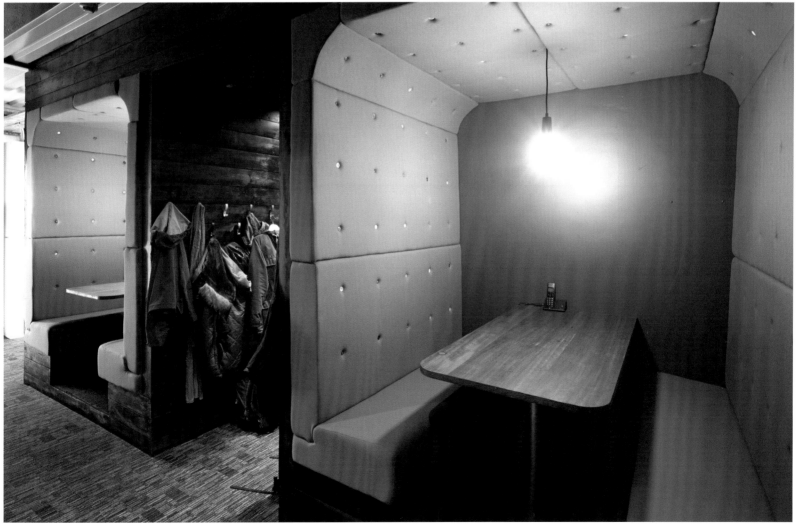

MISCELLANEOUS

Grupo CP Corporate Interior

Grupo CP 公司办公室

Interior Design: Space Arquitectura (Juan Carlos Baumgartner and Gabriel Tellez)

Custom Furniture, Graphic Arts, and Applications: Pentagono Estudio

Collaborators: Marcos Aguilar, Humberto Soto, Enrique Martínez, Diana Casarrubias, Yuri Rodriguez, Antonio Calera, Fernando Lim, Lab Design

Location: Mexico City, Mexico

Construction: GAYA

Furniture: Herman Miller

Photography: Adrenorama & Paul Czitrom

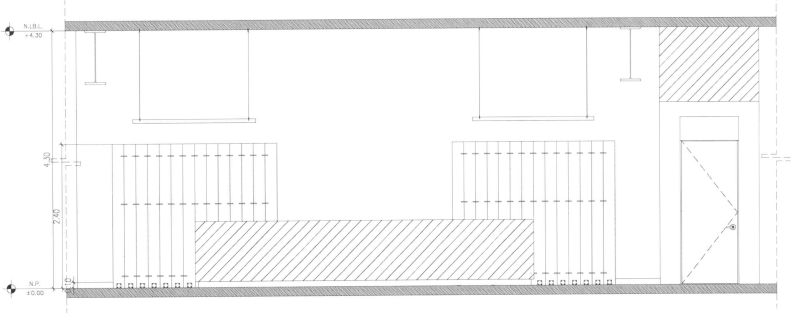

"Terret Polanco" in Mexico City, Mexico, is in the process of LEED certification for its sustainable qualities. Amongst other prominent developments such as "Plaza Carso" and "Museo Soumaya", the skyscraper hopes to become a new environmentally-friendly business hub for the country. Within this structure, Space Arquitectura has constructed the corporate offices of Grupo CP, a project on 3 levels (floors 16 through 18), each comprised of approximately 1,570m².

A railroad on the middle of the floor evokes dynamism, movement, and evolution. This design element was proposed with the objective of spreading these concepts through all company levels.

Specialized furniture, graphic arts, and applications by Pentagono Studio, were generated to evoke an architectural balance in conjunction with the identification of spaces and visual communication. Each level incorporates particular details that generate an integral thematic concept related to the company's business areas: city, health, and automotive.

The shape of the floor plan is trapezoidal with a core towards the back that does not reach the facade. Within this footprint, the architectural program includes

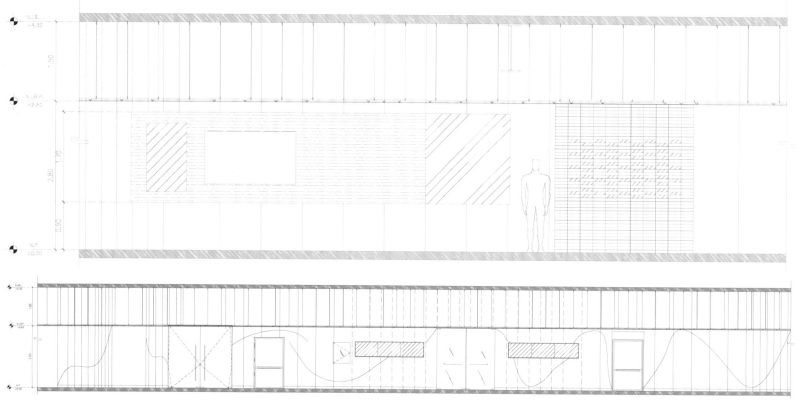

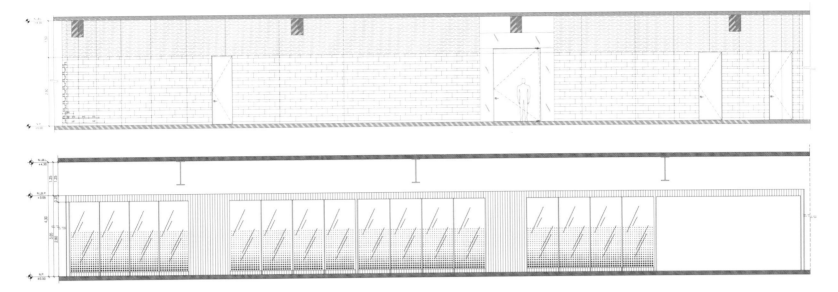

spaces such as a vestibule, lobby, open areas, private rooms, conference rooms, executive dining rooms, informal meeting rooms, one-to-one rooms, training areas, support areas, technology rooms, digitalization rooms, cash register, among others. Level 16 is private with an openness supplying the need for a collaborative environment. Above, level 17 is public with a large training zone. The final floor, level 18, is hierarchically distinguished for its private spaces for shareholders and executives. In the design of these offices, sustainability in operation was contemplated from the start, featuring low energy consumption through lower-than-standard illumination levels, low VOC contents in the products to be used, and a standardization of materials.

墨西哥城"Terret Polanco"具有可持续性特征，正在申请绿色建筑认证（LEED）。本栋大楼与其他出色的房产不同，比如"卡尔索广场"和"索玛雅博物馆"，设计师希望把大楼设计成墨西哥的新环保商业中心。Space Arquitectura 设计公司负责设计了大楼中的 Grupo CP 公司的办公室。办公室分为3层（第16层到第18层），每层楼的面积为 1 570 m²。

地面中央的铁路图案让人联想到动力、运动和进化。设计旨在将这些理念客观地体现在每层楼中。

Pentagono 工作室专门从事家具、平面艺术及其应用，在本项目中主要负责实现空间特征和视觉交流之间的结构平衡。通过在每个楼层加入一些独特的细节，来形成与公司业务领域相关的综合主题：城市、健康和自动化。

空间被规划成梯形布局，核心区虽是设在靠里处，却并未靠墙。在这样的规划方案下，建筑内设有以下空间：门廊、大厅、开放区、私人间、会议室、行政人员餐厅、

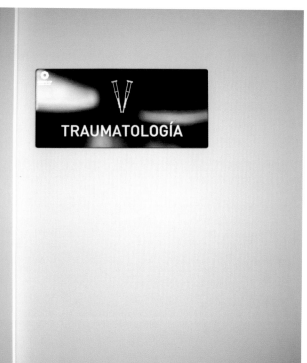

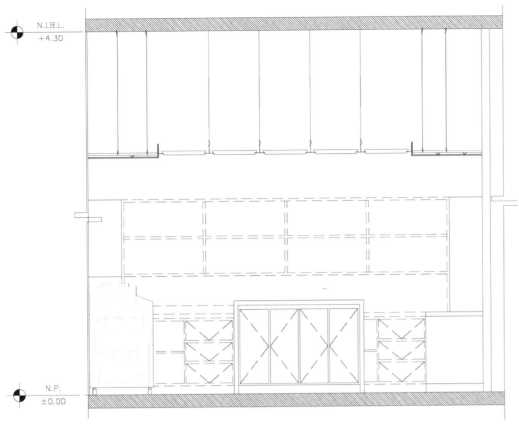

非正式会议室、一对一交流室、培训区、后勤区、科技房、数字化房、收银机等。第16层是封闭空间，仅设有一个开放空间，能满足对合作环境的需求；第17层是公用空间，包括大型培训区；最后是第18层，这一层与其他楼层有等级之分，主要是股东和高管们的私人办公空间。可持续性贯穿了整个设计过程，通过采用比标准用电量更低的照明设备来降低能源消耗，同时选用低挥发性产品以及标准材料。

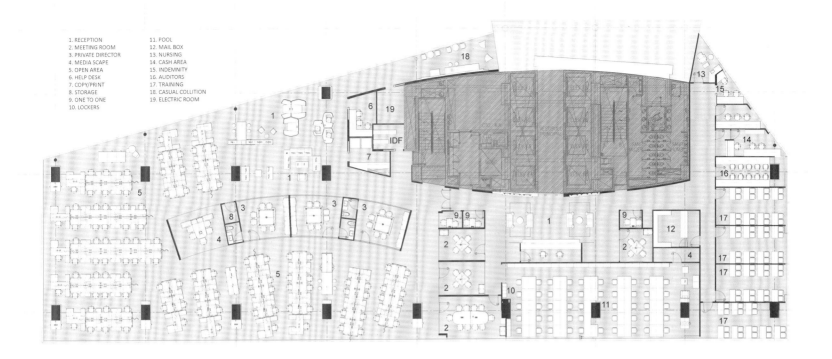

1. RECEPTION
2. MEETING ROOM
3. PRIVATE DIRECTOR
4. MEDIA SCAPE
5. OPEN AREA
6. HELP DESK
7. COPY/PRINT
8. STORAGE
9. ONE TO ONE
10. LOCKERS
11. POOL
12. MAIL BOX
13. NURSING
14. CASH AREA
15. INDEMNITY
16. AUDITORS
17. TRAINING
18. CASUAL COLLITION
19. ELECTRIC ROOM

1. CASUAL COLLITION
2. MEETING ROOM
3. PRIVATE DIRECTOR
4. OPEN AREA
5. TECHNOLOGY ROOM
6. COPY / PRINT
7. STORAGE
8. ELECTRIC ROOM
9. IDF

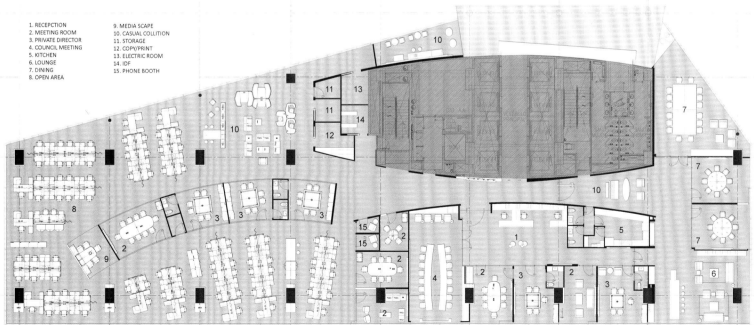

1. RECEPCTION
2. MEETING ROOM
3. PRIVATE DIRECTOR
4. COUNCIL MEETING
5. KITCHEN
6. LOUNGE
7. DINING
8. OPEN AREA
9. MEDIA SCAPE
10. CASUAL COLLITION
11. STORAGE
12. COPY/PRINT
13. ELECTRIC ROOM
14. IDF
15. PHONE BOOTH

MISCELLANEOUS

Silos 13

13号筒仓

Architects: vib architecture

Location: Paris, France

Architects in Charge: Bettina Ballus, Franck Vialet

Area: 4,478.0 m²

Photography: Stéphane Chalmeau, moulinet. R, vib architecture

The project is located 5m away from Paris eastern ring road, at the end of Zac Rive Gauche's large development district. Urban studies conducted by Ateliers LION since 2000, as well as new urban regulation (PLU) updated in 2010, have made room for a new Bruneseau Nord neighborhood. This project is characterized by high rise buildings and mixed programs where architecture and infrastructure meet. To allow for this new development, the city of Paris has asked Ciments Calcia to give up their existing distribution center located near the Seine, and offered a new site closer to the existing rails out of Austerlitz station. Semapa undertook the building of this new project for Ciments Calcia.

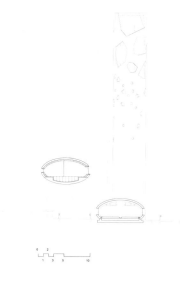
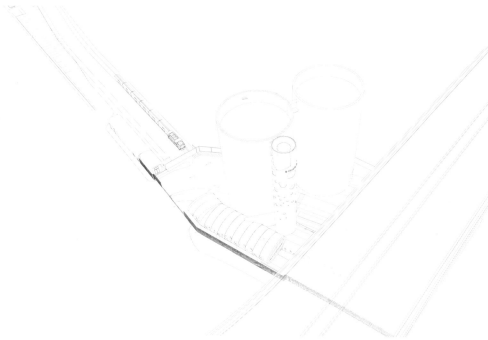

"This project is the first step to develop Paris's new Eastern district. The initial question for us clearly was to insert the project in the coming urban project and bring in "bold design" to the industrial plant. The project was long to design at first, due to high stakes and its noticeable location along Paris' ring road – Europe's busiest freeway with an average 300,000 vehicles a day. The initial 50m high silos project was rejected during the building approval phase – despite urban planning and new regulations allowing for high rises, and we were asked to redesign a 37m high project to fit Paris's usual maximum height. This implied new major constraints. The silos had to be widened to 20m to allow for the

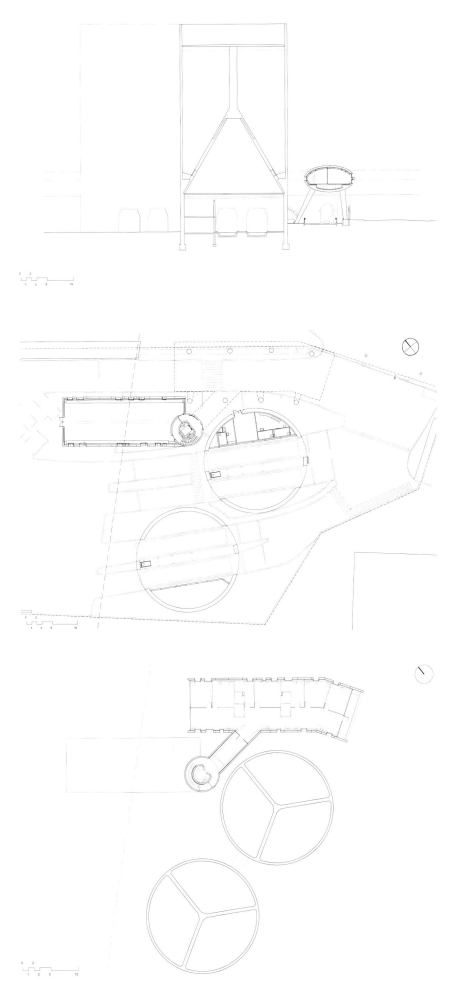
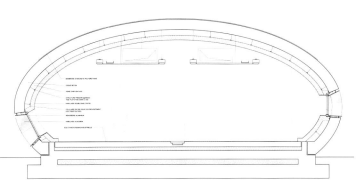
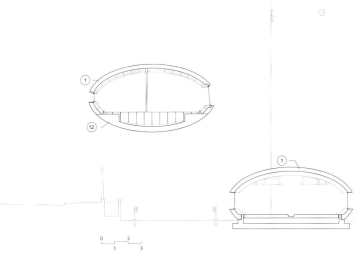

•323•

same volume of cement, making the site almost too small to hold the program."

Floor Plan

Our answer was what it is now: first, the main silos had to be inserted in the one and only position that allowed for all existing flux and networks to work. Then the quality control center, because it had to rest on ground, was slided underneath the "peripherique". The office building was set right on the property limit and high on pillars for trucks to come in below. The whole program could then be linked together by a vertical cylinder housing a stair and lift. The project had become quite simple: the whole program, including offices and quality control center, was inserted in 5 different individual silos, all made of concrete, rising from the ground surfaced with concrete too.

The Scale of the Building

One is now instantaneously overwhelmed by the mass of the project. Although it uses common technical language and fittings, the building has been transformed into a unique work space, dedicated to the material it contains: the silos, the stair tour, the offices, the test center and the ground, are all made from concrete. The material reveals much of its plastic potential. Not one element that stands out. The project is a whole, it is alive, a sort of abstraction of bodies rising from the ground and attracting each other into one unique place.

The shells for the horizontal silos were prefabricated, then trucked-in, lifted, rolled and fitted in no more than 24 hours for each entity.

Polygonal Windows

For the office and quality control center, the light at either end of the "silos" was not enough for all work spaces. The opening on the shells had to be big enough to bring in light but small enough to keep the mass and the surface. The polygonal shape chosen for these openings was imagined as a form of abstraction of 2 different ideas: the shape of the stones that are used in making the concrete as one, or a mineral fragment or a broken rock as another.

The making of the fixed polygonal aluminum windows was complex. The first issue was the geometry of the polygon meeting with the cylindrical shell – that was sorted with the offices 3D tools. The second, more complex issue, was that to do with current regulations that require really sophisticated design for waterproofing and drainage. We called-in Arcora, our usual facade engineers, to help us design these and make them fit current standards.

The vertical lift/stair tower is the most visible piece of the building, which acts as a signal by rising only 5m from the ring road. The design for the openings is similar but they were fitted with a simple inox mesh. We wanted the openings to widen up and lighten up as the tower rises, in contrast to the main silos in the background.

The Fence

Special attention was required for the main fence. It was agreed that the center should be perceivable from the street. Our system allows for visual porosity on the industrial activity, without imposing the heavy trucks onto the new quarter. Our proposal allows for the site to be seen, or hidden, depending on the angle from which one looks.

该项目坐落在 Zac Rive Gauche 地区最大的开发区中，距离伦敦东环路只有 5m。Ateliers LION 从 2000 年开始进行城市研究，并于 2010 年发布了新城市管理条例（PLU），为开发 Bruneseau Nord 新区提供了空间。项目的特点在于建筑的高度以及建筑与基部结构相交形成的复杂结构。为了实现这一新规划，巴黎政府建议 Ciments Calcia 放弃原位于塞纳河岸的配销中心，并给它提供一块靠近奥斯特里茨车站的新场地。色玛帕公司负责了该项目建筑方面的工作。

"该项目是开发巴黎新东区的第一步。"我们所要做的就是大胆设计这个工作，让该项目融入到未来城市规划。由于建筑所在位置十分醒目，位于巴黎环城路边，这条路线是欧洲最繁忙的高速公路，平均每天有 300 000 辆车行经此地，而且项目的风险极高，因此项目经过了长时间的设计。尽管城市规划和新条例准许建设高建筑，然而 50m 高的筒仓项目在建设审批阶段就被驳回，仅允许我们重新设计一栋符合巴黎常规建筑高度限制的 37m 高的项目。这无疑又引出了另一个限制条件，筒仓必须扩大 20m，从而能够容纳同样体积的混凝土结构，这样一来，场地就太小了，甚至很难容纳这个项目。

平面图
我们先了解项目的现状：首先，主仓只有一个位置可选，而且所有的流通网都必须流畅。由于质量控制中心必须设在地面上，因此设在了环城大道的下方。办公建筑恰好落在场地的右侧边缘处，其结构被柱子托起，下方允许车辆行驶。楼梯和电梯所在的立式圆柱形建筑将所有的建筑连接起来，使项目显得十分简单。整个项目包括办公空间和质量控制中心，分别设在 5 栋独立的筒仓中。这些筒状的建筑都是用混凝土建成的，建筑所在的场地也是水泥地。

建筑规模
建筑规模之大，让人在顷刻间便被征服。虽然建筑在技术和配置上都略显普通，但是建筑所采用的材料使建筑成为了一个独特的工作空间：筒仓结构、楼梯、办公室、检测中心和场地都是用混凝土建成。这种材料充分发挥了自身的可塑性。项目作为一个整体，没有元素特别突出，它拔地而起，就像是抽象的生命体，建筑与建筑之间相互吸引，形成了这个独特的空间。

水平延伸的筒状壳式结构是预先制造好的，然后运到场地内，经过抬升，翻滚到合适的角度，再进行安装而成，每个结构所花的时间都超过 24 小时。

多边形窗户

由于办公空间和质控中心的两侧都不能给所有的工作空间提供充足的光线。因此，壳式结构必须设计大小合适的窗口，这样既能保证室内的光线，又能维持结构的规模和外观。多边形的窗户有两种抽象的含义：组成混凝土结构的石子的形状，以及矿物碎片或碎石。

固定多边形铝窗的制作过程十分复杂。首先，多边形的几何结构必须与圆柱形的外壳相匹配。这一问题的解决得益于办公室 3D 工具的使用。其次，现行规定要求设计高级的防水和排水系统。因此，经常与我们合作的 Arcora 的外观工程师们也被邀请来解决这些问题，使它们符合现行标准。

竖立的电梯和楼梯塔是这些建筑中最醒目的设计，它仅仅比环城路高出 5m，因此也是最具标志性的设计。所有窗户的设计手法都很相似，都装有简单的不锈钢网。随着塔楼高度的增加，开口也越来越大，空间也更加明亮，与背后的主仓形成对比。

栅栏设计

本项目格外注重主栅栏的设计。为了保证从道路上可以观看到质控中心，我们采用了视线孔隙度恰到好处的栅栏，人们可以透过栅栏看到工业活动，而从室内又不会看到户外繁忙的车辆。这种栅栏既能隐藏空间，也能把空间向人们开放，主要取决于人们观看的角度。

第一部分
可持续项目

"因为该地将实施新城市规划，所以如果我们想在 Bruneseau 继续发展，现有的工厂必须移除，"法国 Ciments Calcia 配送中心总经理 Jerome Lestringant 说道。"然而托比亚克原配送中心具有密集性：80% 的客户都在场地周围 30 km 范围以内。显然，我们想维持这种距离。"

尽量靠近城区

"原托比亚克配送中心位于城市中心，其位置具有战略性意义。项目采用的水泥

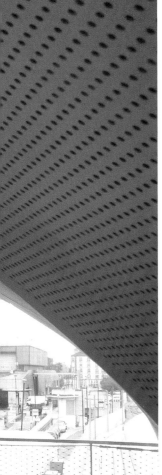

Section 1
A Sustainable Project

"Considering the new urban project, it was obvious that our existing plant had to be moved if we wanted to keep our activity running in Bruneseau," says Jerome Lestringant, general manager for Ciments Calcia's distribution centers in France. "The catchment area of our existing center in Tolbiac was quite concentrated and dense: around 80% of our clients are within 30 km around the site. We obviously wanted to stay near them."

Stay as close as possible to the City
"The key quality for the existing plant in Tolbiac rests in its strategic position at the heart of the city. While 100% of the cement comes upstream by train, the last few kilometers made downstream by truck by our clients, is usually done in the opposite direction from main urban traffic directions. Moving the plant away from this position would have increase truck movements by 15,000 a year, in the direction identical to the already oversaturated general traffic."

Section 2
"Our aim was to create a sustainable project," says M. Lestringant. "We wanted to maintain the rail delivery, and to keep the possibility to supply our cement via the nearby Seine. That was another incentive to stay close to our old site. The city of Paris heard our arguments and approved our plans. They rightly considered that our proposal fitted with their vision of preserving mixed uses in the new Bruneseau district."

A piece of art
A work of art, imagined by Laurent Grasso is currently being tested. It will give yet another dimension to this already atypical urban project.

大部分都是供应商用火车运来的，而其他则是由客户用卡车提供的，避开了城区的主要交通流通线。如果把工厂从原址搬走，将给已经饱和的交通线每年增加 15 000 的车流量。"

第二部分
"我们的目标是建造一个可持续项目，" M. Lestringant 说道。"我们想保持铁路运输量，尽可能地从附近的塞纳河地区获取水泥。这也是我们坚持选择靠近原址的原因之一。巴黎政府听取了我们的提议，并批准了项目计划。他们认为我们的提议是与在新区中保留多用途建筑的远见是一致的。"

艺术品
这项由 Laurent Grasso 构思的艺术品正在接受考验。它将开创非典型都市建筑的新领域。

Talent Garden office, Brescia - Italy

意大利布雷西亚 Talent Garden 办公室

Contributor: Talent Garden

Photography: DAVIDE D'AMBRA

"Passion working space" is the green digital co-working space.

The charming mix of co-working space and green garden encourages creativity, collaboration, supports recycling and eco-friendly design.

The name Talent Garden was the major basis for the whole interior design - digital talents space with 100% recyclable cardboard architecture.

The campuses have a set of a careful research based products, such as cardboard desks with internal reinforcement in wood, relax areas with eco armchairs and chaise-lounges made out of the recycled paper.

The campus has been further enriched with the green areas made out of the environment-friendly lawns that give a feeling of a real grass touch. Even the luminaries/lamps are made using natural materials, such as Japanese paper that is produced from natural components, which react to fluctuations.

The message behind Tag design is not only to have a nice space but also to foster our responsible behavior and respect for the future. We live in a very fragile ecosystem that has to be taken care of.

这个绿色数字联合办公空间正是一个"充满激情的工作空间"。

公共办公空间与绿色的花园的结合，形成了这个鼓励创新、协作、循环利用和环境友好的设计，十分迷人。

整个室内设计是以公司名称 Talent Garden 为基础的，从而形成了这个以 100% 可回收纸板为结构的数字人才空间。

空间内设有一系列经过精心研究的产品，例如，木构架纸板桌子，休息区中的环保扶手椅，还有利用回收纸做成的长沙发。

另外，还有用环保草坪打造的绿色区域，这不仅丰富了空间，还给人一种真实草坪的感觉。甚至灯具也是采用天然材料做成，例如用天然成分做成的日本纸，能对波动产生反应。

因为我们居住在一个需要呵护的脆弱的生态系统中，因此品牌设计所蕴含的不仅是设计"美观的场所"，还包括培养对我们自己行为负责的责任心，以及对未来的尊重。

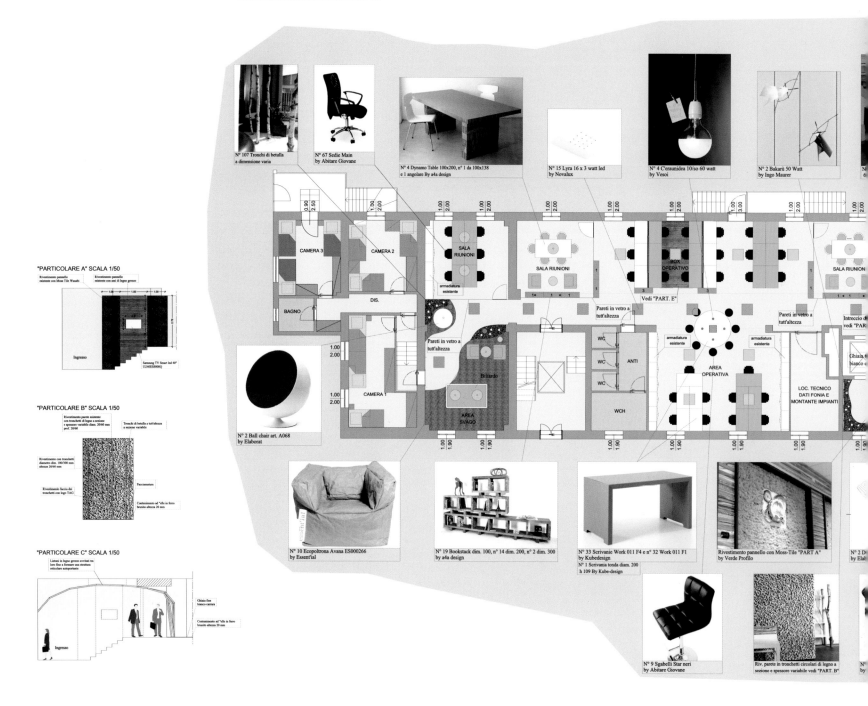

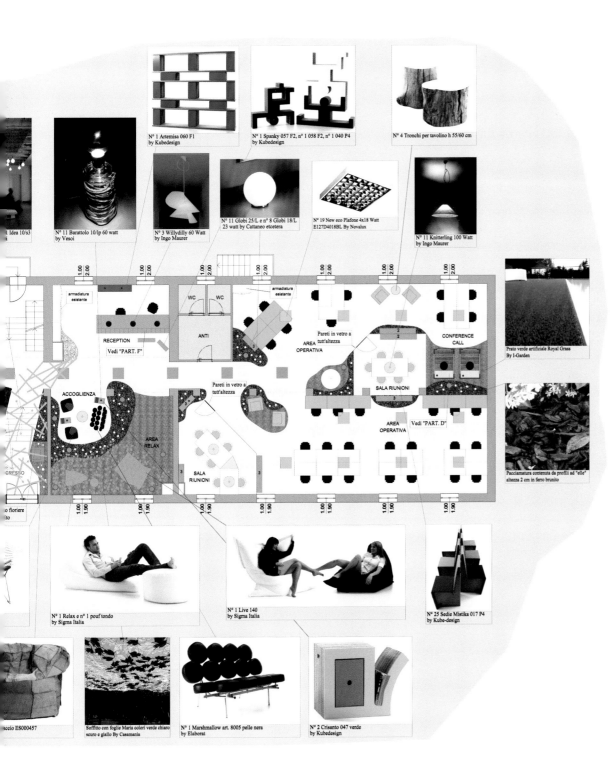
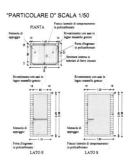
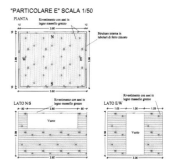
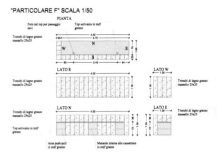

Johnson & Johnson Eyesight Health Institute

强生视力健康研究所

Architects and Design Agency: Sergey Estrinm Architectural Studio

Architects: Sergey Estrin, Sheremetyeva Julia, Tatiana Kulikova, Tatiana Grishina

Client: Johnson & Johnson

The Vision Care Institute - Training Center for Professionals (Vision Correction)

A division of Johnson & Johnson vision Care is known worldwide as a manufacturer of soft contact lenses ACUVUE.

The office is located in a new multifunctional complex next to Yakimanka street with views on the Crimean Bridge and the Church of St. John the Warrior.

The most important element of the interior is the reception of Corian with "eyelashes" of glowing acrylic glass. All items are made in a single copy according to the drawings of architects. Eyelashes have sectional signs "+" and "-" as the diopter lenses.

Near the reception intersect 4 corridors leading into different zones: administration, auditorium and cafeteria and offices to check the eyes.

Color palette: monochrome with colorful accents.

Dynamic space with different texture and broken lines.

The walls are decorated with paintings of contemporary Russian artists.

视力保健研究所——专业人员培训中心（视力矫正）

强生的下设部门强生视力保健机构是世界知名软形隐形眼镜"安视优"（ACUVUE）生产商。

新办公室坐落在Yakimanka街边的综合大楼中，能远眺克里米亚大桥和圣约翰战士教堂。

室内设计中最重要的元素是可丽耐大理石接待台，以及醒目的有机玻璃背板，上方还有"睫毛"字样。所有的装饰都出自建筑师的设计图，独一无二。"睫毛"字样上有区别性的标志"+"和"-"，代表着屈光镜。

接待台附近交叉的4条过道，通往不同的分区：行政区、会议区、餐厅区和眼睛检查区。

色彩方案：单色为主，彩色为辅。

由折线组合而成的活力空间，结构与众不同。

墙壁则用俄罗斯当代艺术家的作品来装饰。

EB Group Showroom & Office

EB 集团展厅和办公室

Design Agency: plajer & franz studio

Location: Berlin, Germany

Here, round, fluid forms mixed with high-quality materials and a restrained colour palette create an inviting atmosphere whilst clearly expressing competence and trust at the same time. The design of the "showroom" for the EB GROUP respects all the needs and functional assets of an open customer service centre. plajer & franz studio also designed the office spaces and a private apartment floor for the Berlin headquarters of this international real estate company.

圆形的结构、流畅的布局、高质量材料与朴素的色彩相结合，共同组成了一个迷人的空间，清晰地展现了集团的能力和信誉。EB集团展厅的设计具备了一个开放式客服中心所应有的一切功能，内部设备齐全，能满足任何使用需求。plajer & franz设计工作室还为该集团设计了办公空间，另外，还有专门为柏林国际房地产总部设计的私人公寓。

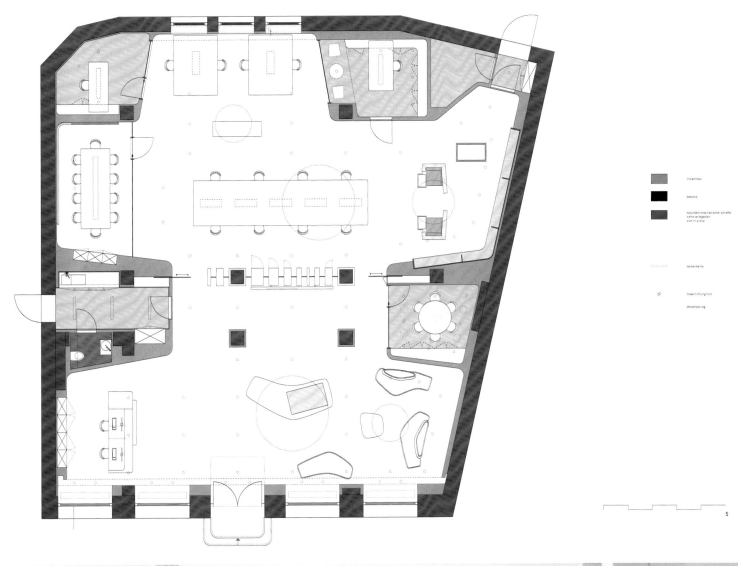

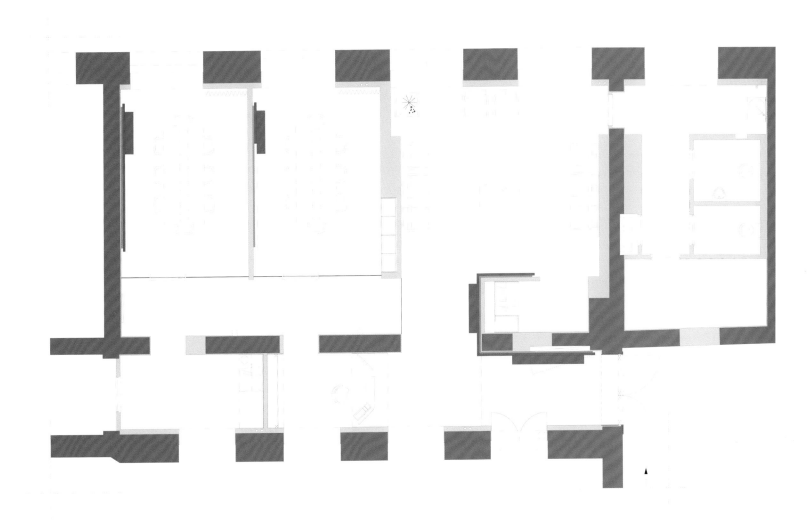

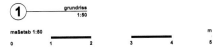

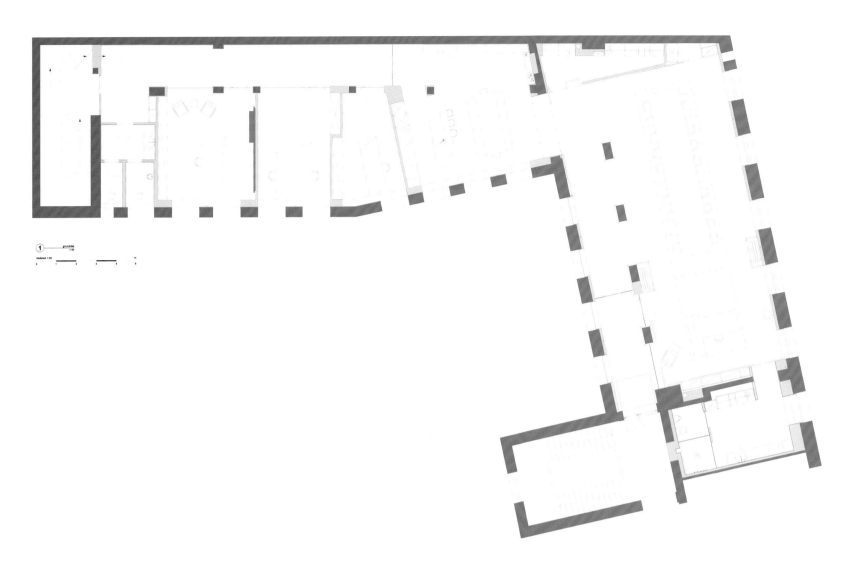

NON-PROFIT/GOVERNMENT

Headquarters of the Fondation Jérôme Pathé Seydoux

百代基金会总部

Architects: Renzo Piano Building Workshop

Design Team: B. Plattner and T.Sahlmann (partner and associate in charge) with G.Bianchi (partner), A.Pachiaudi, S.Becchi, T.Kamp; S.Moreau, E.Ntourlias, O.Aubert, C.Colson, Y.Kyrkos (models)

Total Building Area: 2,200 m^2

Location: Paris, France

Client: Fondation Jérôme Seydoux - Pathé

Photography: Paul Raftery, Michel Denancé, RPBW

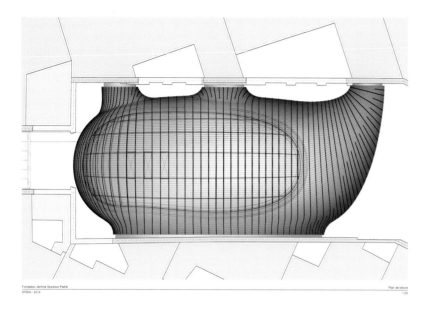
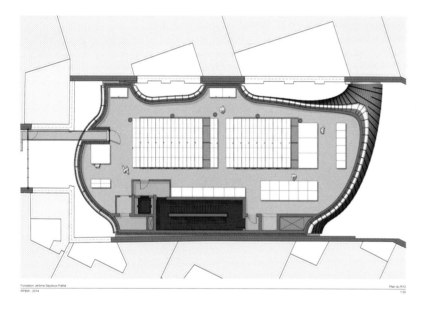
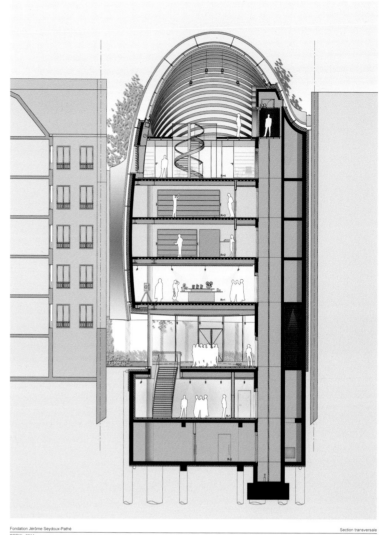
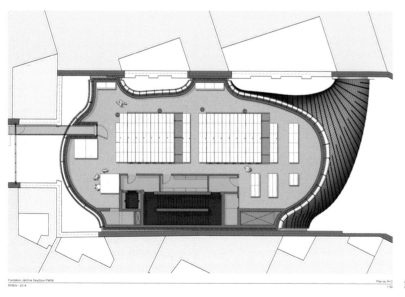
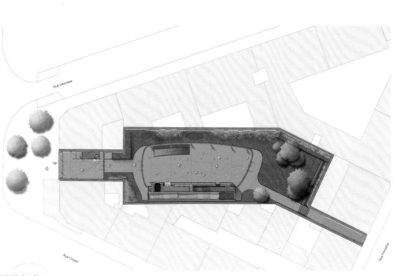

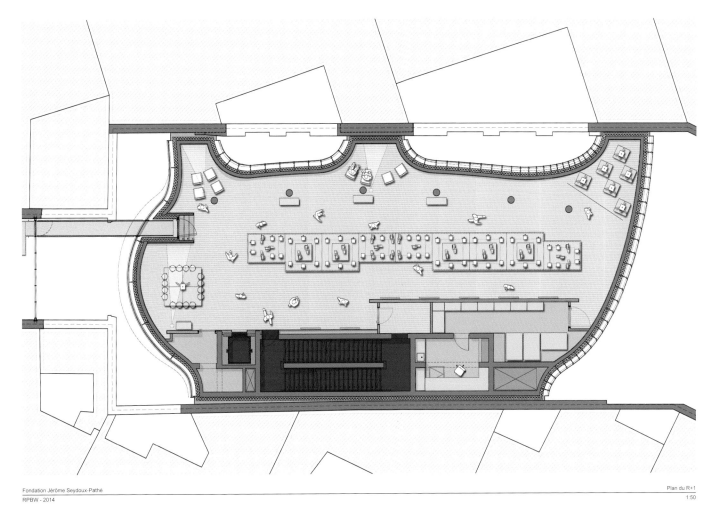

Fondation Jérôme Seydoux-Pathé
RPBW - 2014
Plan du R+1
1:50

The Fondation Jerôme Seydox-Pathé is an organization dedicated to the preservation of Pathé's heritage, and to the promotion of the cinematographic art. Its new headquarters will be located in avenue des Gobelins, on the site of a 19th century disused theatre.

The new building will house Pathé's archives, some exhibition spaces related to the cinematographic art, including a 40-seat screening room, and the offices of the foundation.

The project calls for the demolition of the 2 existing buildings to create a more organic space that better responds to the restrictions of the site.

The facade on the avenue des Gobelins will be restored and preserved, due to its historical and artistic value. Decorated with sculptures by Rodin, it is not only a historical landmark, but also an iconic building for the Gobelins area.

A new transparent building just behind the facade functions as the foundation's public access. Looking like a greenhouse, it offers a view on the interior garden through the basement of the new egg-shaped building that houses the project's main functions.

The peculiar design of this 26m high building is determined by the site's major limits and requirements. In particular, it respects the distances with the adjoining buildings, while at the same time creating a new space for an interior garden.

The glazed form of the building is only perceived from the street through the and over the restored facade like a discreet presence during the daytime, while softly glowing at night.

百代电影遗产保护组织（百代基金会）致力于保存百代电影公司的遗产，推动电影艺术的发展。其新总部将设在高伯兰大街，替代原先废弃了的19世纪电影院。

新建筑将设立百代档案室和展示电影艺术的陈列空间，比如40座放映厅和基金会的办公区。

该项目要求拆除原来的2栋建筑，来获取更为有机的空间，从而能够更好地应对场地的限制。

朝向高伯兰大街的建筑被罗丹创作的雕塑装饰着，它不仅是历史遗迹，而且充当了整个高伯兰地区的标志性建筑的角色，具有历史和艺术价值，因此其外表将经过修复而保留下来。

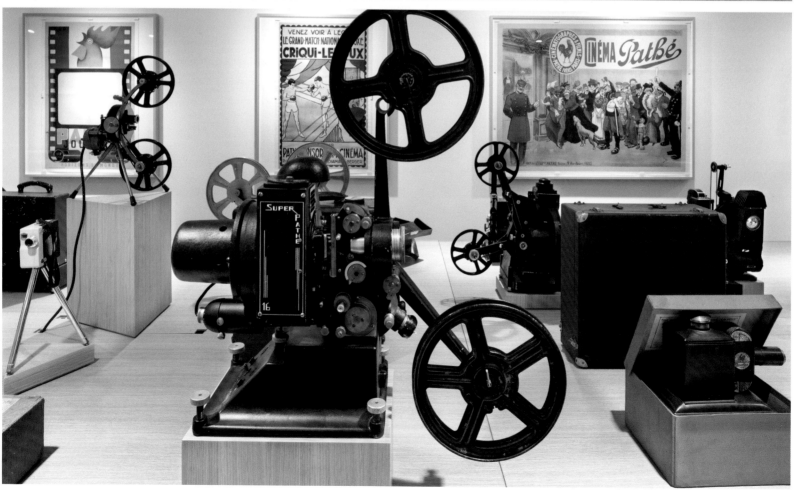

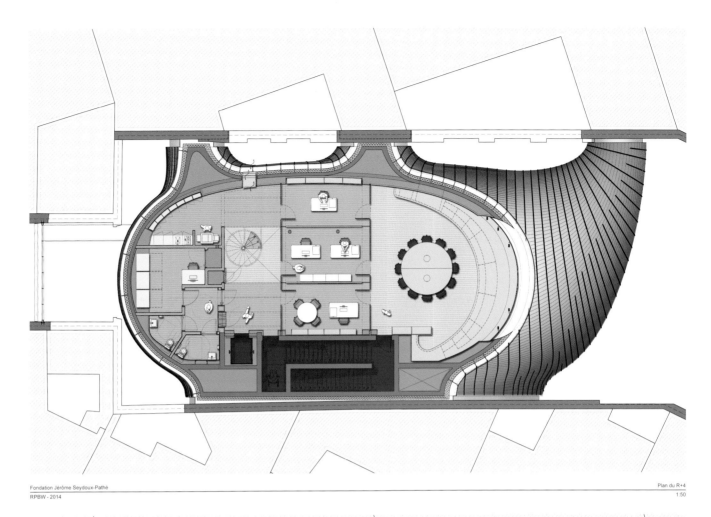

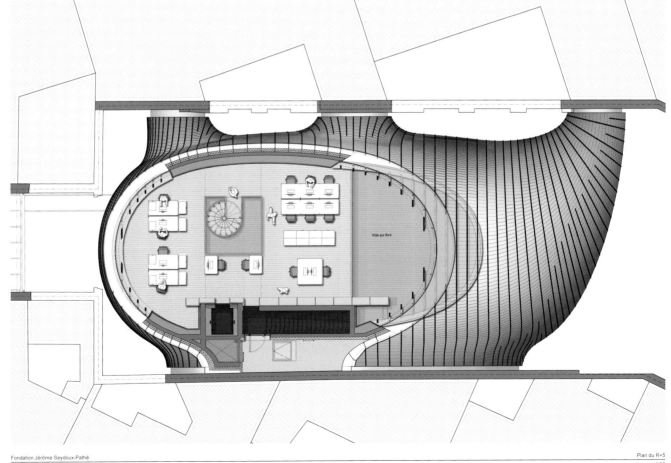

在这道外墙后面是透明的建筑，充当了总部的公共区域。蛋壳状的建筑里被室内花园装饰着，恰似一个温室。该项目包含的主要功能区都设在这栋建筑里。

建筑高26m，受其场地的限制和要求，建筑的设计十分独特，既与相邻的建筑保持了一定的距离，又给室内花园提供了新空间。

这栋被玻璃覆盖的建筑只有从街道的旧建筑大门处才能看到，白天显得十分庄严，夜晚却闪闪发光，安静而又生动。

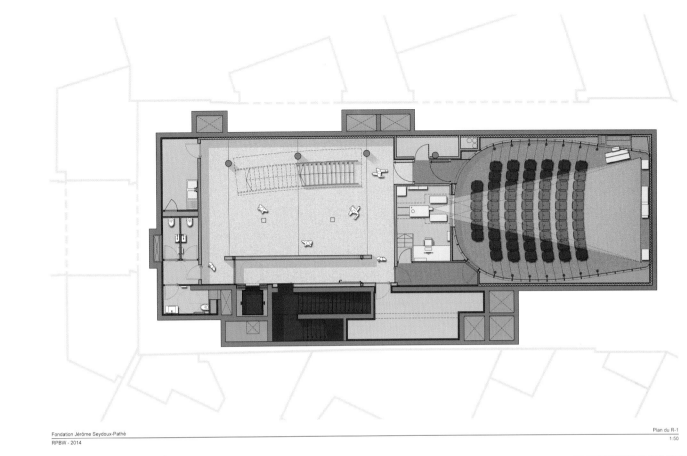

Fondation Jérôme Seydoux-Pathé
RPBW - 2014

Plan du R-1
1:50

NON-PROFIT/GOVERNMENT

NS Stations

NS Stations 办公室

Architects: NL Architects (Pieter Bannenberg, Walter van Dijk, Kamiel Klaasse)

Client: NS Stations

Area: 7,000 m²

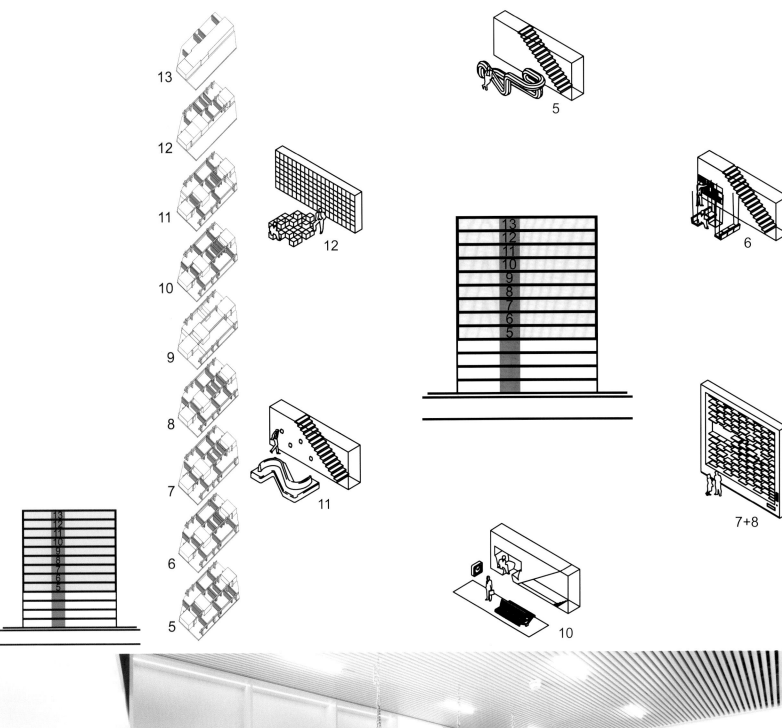

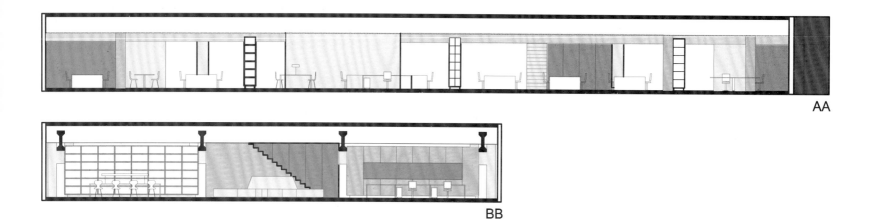

AA

BB

Competition organized by Spoorbouwmeester Koen van Velsen, 1st prize.

The Katreinetoren in Utrecht is the home base of NS Stations, the department of the Dutch Railway (NS) that develops and services railway stations in the Netherlands.

The 15-storey 55m high tower is built right on top of Utrecht Centraal, the biggest railway station in the Netherlands. The entrance is positioned conveniently in the Central Hall!

The width of the building is based on the dimensions of the railway tracks below. The interior is dominated by this concrete structure: the elevator core and structural beams remain in sight.

A reconstruction took place in 1999 when the Brutalist mid-seventies building was wrapped in a glass skin.

The current floor plan consists of typical cellular offices along the facade and an oversized hallway in the center.

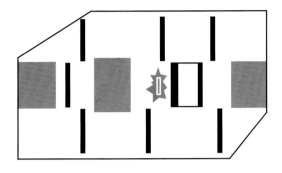
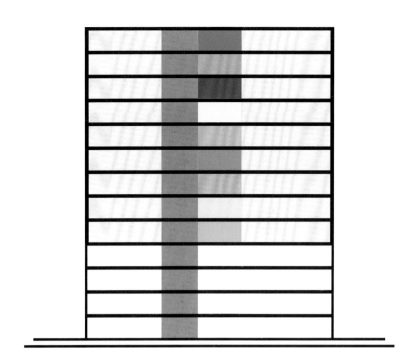

special element in entry zone   each floor different entry zone

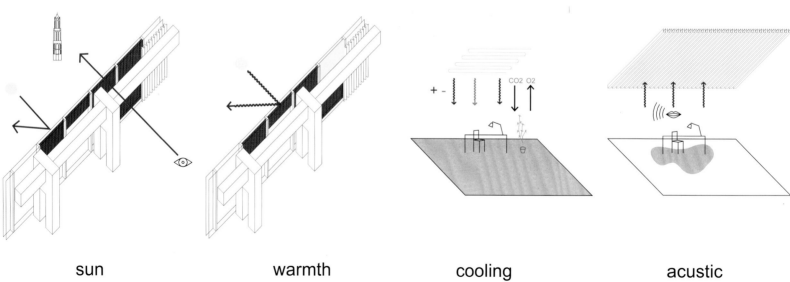

sun　　　　　　　　warmth　　　　　　　　cooling　　　　　　　　acustic

本项目在 Spoorbouwmeester Koen van Velsen 组织的比赛中荣获一等奖。

NS Stations 是隶属于荷兰铁路的一个部门，其本部设在荷兰乌特勒支的 Katreinetoren 建筑中。荷兰铁路主要负责荷兰的火车站建设和车站服务。

这栋大厦共有 15 层，高 55m，正好建在荷兰最大的车站乌特勒支中心车站上，入口设在中央大厅中，十分便利。

建筑的宽度与下方的火车轨道的规模一致。内部主要是裸露的混凝土结构，电梯间和清晰可见的建筑横梁。

1999 年，建筑经过了翻修，当时盛行 17 世纪中期"野兽派"玻璃建筑。

如今的楼层规划中包含了典型的移动办公空间和宽阔的中央走廊。

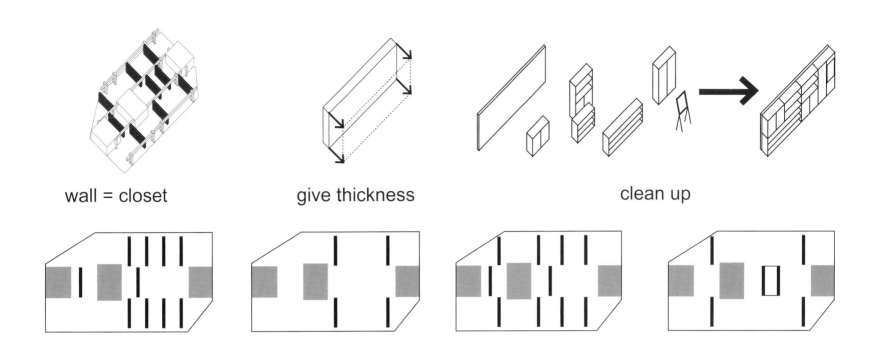

wall = closet · give thickness · clean up

different positions and density of walls

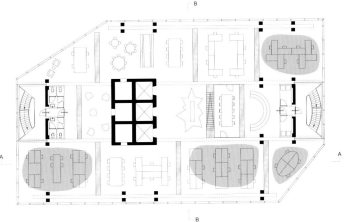
clusters

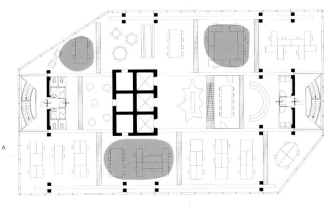
flexworking

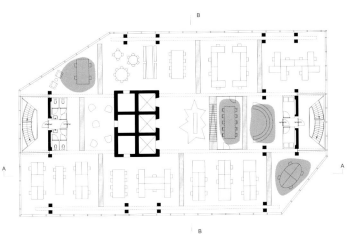
conference

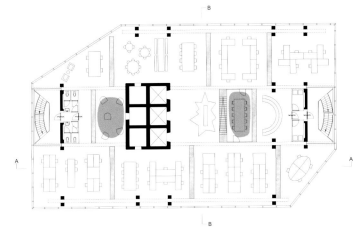
private conference

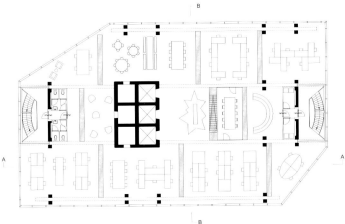
lunch area

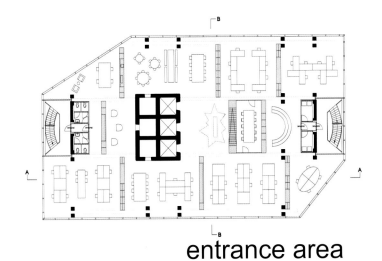
entrance area

设计中主要引入了一系列可以组织空间的墙壁。通过加宽隔墙，使它成为储物空间。墙壁还充当了陈列架、书架、活动板、档案存放区、公告板，甚至是楼梯，减少了空间的凌乱感，使空间更加整洁有序。

功能不同的厚墙，形状各不一样，营造出来的氛围也大相径庭。

设有厚墙的开放空间被划分成几个大小不同的区域，与普通的走廊连接房屋的布局恰好相反，形成了一个连续流畅的空间。空间未专门设计交通空间：总值＝净值。

这种规划带来的财富是连通性的提升。自然的社交空间取代了划分成各种相对独立的小间的办公空间。

这些墙挖掘出了建筑空间的潜力，创造出巨大的空间，同时也保持了空间的紧凑感和亲切感。

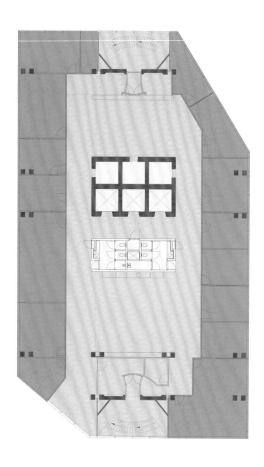
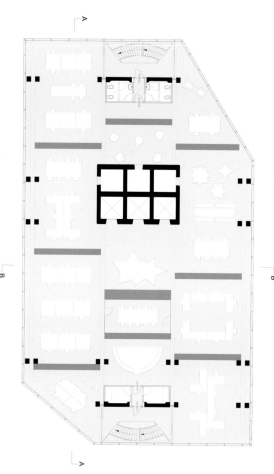

The basic idea is to introduce a series of perpendicular walls that will organize the space. By "inflating" these partitioning walls they can become storage spaces.

The shelves can absorb most of the clutter that normally spreads trough the office, such as wardrobes, bookshelves, flip boards, archive, bulletin boards, and sometimes even the stairs. The Thick Walls™ will clean up the space.

The Thick Walls™ can take different shapes to introduce specific functionalities and atmospheres.

The open plan with the Thick Walls™ divides the space in several clusters of different sizes. As opposed to a regular layout with cells linked to a corridor, a continuous, flowing space is created. Traffic space is eliminated: gross becomes net!

The main asset of this system is increased connectivity; instead of a workforce relatively isolated in cells, here interaction will come natural.

The perpendicular Wall Unit explores the spatial potential of the building; an extensive "field" condition is created while keeping a sense of compactness and intimacy.

In the near future NS Stations plans to introduce the so-called flex-work system which means no more

fixed workstations but a range of specific working environments to choose from. The proposed layout will be able to accommodate this intricate workflow, but for the time being most of the clusters will be organized as fixed working stations.

A working station consists of an ergonomic office chair and a table that is individually adjustable in height. With this unit an endless number of configurations is possible.

The dividing Wall Units can be placed in different 'densities' to create fitting floor plans for the various departments.

A series of workshops with the future users will be organized to find a specific disposition for each floor.

The 9 levels of 700m² each will be arranged according to the same principle but all will be configured slightly different.

Each floor will be introduced by a "special": a singular piece of furniture located at the elevator core that represents the different departments. This "conversation piece" is a kind of welcoming gesture: perhaps a logo-shaped bench, a swinging multifunctional lounge chair, a 2-storey high candy machine, a sculptural reading table spelling NS, a Happy Hour clock maybe...

在不久的将来，NS Stations 将引进所谓的弹性工作体系，这就意味着固定的工作将被各种可以选择的工作环境所取代。此次提出的布局将能够适应这种复杂的流通方式，但目前各种空间还是会按固定的工作台来布局。

这些工作台是由符合人体工程学的办公桌椅组成，而且其高度都是可以单独调节的。在这样的布置之下，再配上无数的设施也都是可以的。根据楼层部门的设置情况，这些隔墙设置的数量也不一样。

为未来使用者设计的一系列工作区将会设在每层的最佳位置上。

面积为 700m² 的那 9 层楼将以同样的原则来设计，当然也会略有不同。

每层楼将设计一个"特殊空间"：在电梯区中设计了指示牌，标明了不同的部门。交谈区的家具具有一种热情的姿态：如，与标志形状一致的长凳；多功能的摇椅；两倍高的糖果机；NS 形大阅读桌，以及"快乐时光站"等。

NON-PROFIT/GOVERNMENT

Northern Territory

泰利驿站

Design Agency: Beijing Qinshi Architectural Design

Designers: Li Yiming, Lv Xiang

Northern Territory, Dongsheng Technology Park, Zhongguancun, as a creative social and economic organization, provides research, manufacture and operation sites, communication, network and office shared facilities, and systematic training and inquiring services. It also supports policy, financing, law and marketing etc. so as to decrease risks and costs of startup enterprises and increase their surviving and success rate.

Northern Territory is on the south side of the first floor in creation center office building of the campus, which features 3 areas. The main functional spaces include individual office and team open office alike open rental offices, and others relatively separate public spaces, such as enterprises incubating offices, meeting rooms, reception space, discussion rooms, library and presentation rooms etc. what's more, there are closed offices for the whole serving and supporting team.

The project aims at reflecting the intrinsic attributes of space, echoing with its creator to generate more passion and inspiration that are always come from the yearning for freedom, while care for creation and growth is Taili Stage's destiny. Therefore, "freedom" and "care" are the best explanation to the essence of this case.

At first, we segregate the space into south and north areas by drawing a free flow curve through the whole space. The curve extends and closes to form various relatively closed oval spaces which are used as public spaces, including meeting rooms and reception rooms and so on. The southwest side is more centralized and closed enterprises incubating offices; while the east side of the main entrance is logistics support areas that serve the interior and exterior space more conveniently because it is separated in some degree.

Especially, a 7-storey high volume space is kept as the entrance hall without any real function. A 2-storey high half closed arch-shape form presents its protection to the entrepreneurs, which also means the great expectation and freedom.

Carpet design is a feature in this case. From the entrance to the open working area are covered by different width and color striped carpets, which become many colorful scenery lines, thus break the traditional carpeting and color matching methods. In such a way, the essence of creation is displayed, so does an iridescent "T" shape stage. All these bless each entrepreneur successes on their way.

The main color palette of each facade and ceiling is purely white, echoing with the carpet. The ceiling seems like a "honeycomb" with an irregular polygonal tensile membrane structure as main lighting set that shines the whole space with its soft light, and gives a sense of modern and warm.

Concaved arc-shaped desk and wall panel guides visitors in the space. Curved openings on a curved wall form a free hyperbola, which embodies freedom, passion, dream and technology. The ceiling above an open working area is designed on honeycomb hexagon elements. Then hexagons transformed into various polygons that combine to an irregular polygonal suspended tensile membrane, which inspires people's thinking, as well as curiosity and desire to explore.

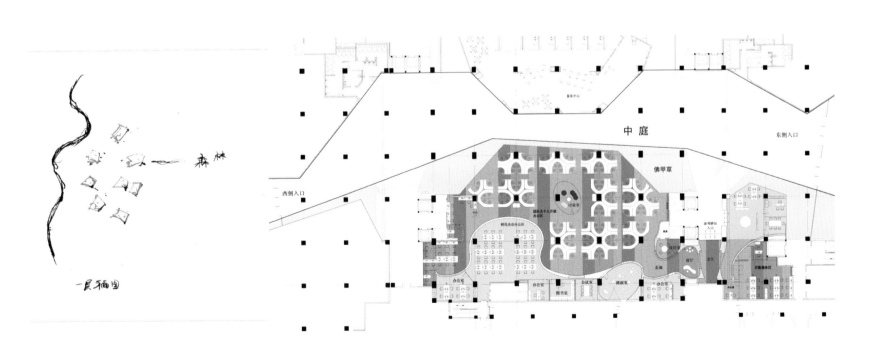

Columns are with shading colors from dark to light, finally become to be white and integrate with the ceiling, similar to a big tree that shelters those entrepreneurs. The office furniture forms a "U" letter surrounding each column, which are also operable for big or small teams. They are roots of the big tree, deep rooted into the colorful stage.

Inserted among the open working area, presentation room and the discussion room, a small indented area is carefully designed, which not only ensures the continuity of the outer space, but also functions as a transition between various spaces, visually and mentally.

The chairs in the discussion room are bright in color, warm and cold color palette highlights a free and opponent ambience during a debate. Of course, each participant can select their own seat according to their own preferences. However, the desk top and all around are white, thus soften the atmosphere.

The presentation room is a special-shaped space, because different occasions have different layout. White walls help each presentation, and linear pattern carpets are paved linearly to give a sense of woven feeling. What's more, the various sized colorful circular mould chairs break tradition and add more creativity and freedom to the space. The ceiling design is based on "butterfly" elements. Several abstract flying butterfly shapes means these startups break out from a cocoon and take off after they succeed in the presentation.

In the center of the open working area is an elliptic multifunctional space which can be used as reception area, discussion room, meeting room and lounge room etc. transparent glasses without reaching the ceiling are used to separate this place, which fully displays its open and public character. The colorful modern furniture integrates easily with the colorized carpet, thus adds more transparency to the space.

Also, we are inspired by the 2014 World Cup, Brazil. Football net, as a segregating material and form, is used to separate various spaces, in such a way, the space is full of sporting atmosphere, also no lacking of a soft touch. We believe that entrepreneurs are more passionate during the World Cup. A transparent glass screen separates the office from the public hall area. Colorful fluid spaces are lighted with soft lights that permeate to the hall atrium, and telling moving stories about freedom and creation.

中关村东升科技园·泰利驿站是一种创新型的社会经济组织，它通过提供研究、生产、经营的场地，通讯、网络与办公等方面的共享设施，系统的培训和咨询，政策、融资、法律和市场推广等方面的支持，降低创业企业的风险和成本，提高企业成活率和成功率。

泰利驿站位于园区创新中心办公楼的一层南侧。主要功能包括个人办公区，团队办公区等开敞式出租办公区域，孵化企业办公区、会议室、接待室、讨论室、图书室、路演室等相对独立的公共空间，以及为驿站提供整体服务的后勤支持团队封闭式办公区等3大区域。

如何诠释出空间的内在属性、与创业者产生共鸣、使创业者迸发出更多的激情与灵感，是本案设计的核心所在。激情与灵感往往来源于对自由的向往，而对创新、对成长的呵护是驿站的使命。因此，"自由""呵护"是对本案内在属性最好的诠释。

首先，我们用一条自由的曲线贯穿整个空间，将空间分隔成为南北两个区域，

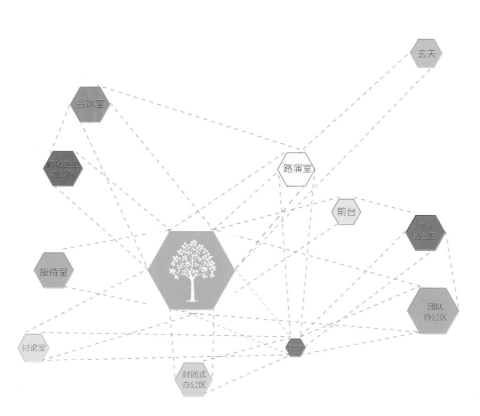
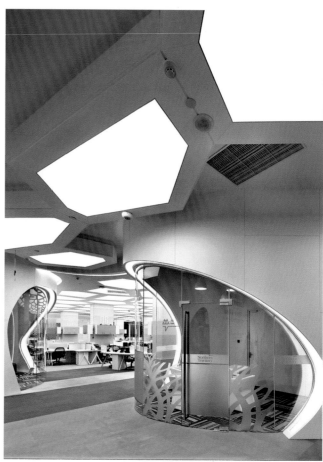

随着曲线的延展与闭合，形成各种卵形的空间，作为会议室、接待室等相对封闭的公共空间。曲线的北侧主要为个人及团队的开敞式办公区域，靠近大堂中庭，采光良好。西南侧为孵化企业办公区，相对集中、封闭。主入口的东侧为后勤支持办公区，靠近主入口，相对独立，利于对内对外服务。

在入口处，我们特意保留了一个高达7层的挑空空间作为玄关，没有赋予任何实质的功能，只是做了2层高的弧形半围合形态，想要传递出对创业者的呵护、成长的巨大空间与自由这个寓意。

地毯的设计是本案的一大特点。从入口到整个开敞办公区，我们均采用了不同宽度、不同颜色的条形地毯进行铺装，形成一条条多彩的风景线，突破了传统的地毯铺装及色彩搭配方式，体现出创新的本质，也象征着绚烂多彩的"T"形舞台，祝愿每个创业者都能找到自己的成功之路。

与多彩的地毯相对应，立面与天花板基本都采用纯净的白色作为主色调。天花板采用类似"蜂巢"的造型，不规则的多边形张拉膜作为照明主体，使整个空间散发着柔和的光晕，同时传达出现代与温馨之感。

内凹的弧形前台与背景墙，给人们带来了明确的路线指引，在弧形的墙面上再配以弧形的洞口，形成一条条自由的双曲线，表达出自由、热情、梦幻、科技。开敞办公区域的天花板采用蜂窝六边形元素为基础进行设计。六边形转变为不同的多边形，然后将其自由组合成不规则的多边形张拉膜吊顶，使人们充满新鲜感和探索欲望，并激发人们的思维。

柱子采用由深到浅的渐进颜色，最终以白色融入天花板，似一棵棵粗壮的大树，全力呵护着创业者。办公家具以"U"形为一组，以每个柱子为中心环形布置，可以相互组合成大大小小的团队，同时，也像是每棵大树的树根，扎根在多彩的舞台上。

在开敞办公区与路演室、讨论室等相对封闭的空间之间，我们精心设计了一个小小的内凹区域，既保证了外部空间的连贯性，又是一个不同类型空间的转换区，实现了视觉上和心理上的平缓过渡。

讨论室的座椅选用了鲜艳的颜色，并冷暖搭配，渲染出讨论时自由、紧张的气氛，每个参与讨论的人也可以根据自己的喜好进行选位。而桌面及四周均为白色，在一定程度上可起到缓和气氛的调节作用。

路演室为一个异形的空间，不同的场合有不同的布置方式，四周的墙面均采用白色，便于路演时的讲演。地毯选用了具有编织感的线形图案及铺装方式，椅子也打破常规，采用了大大小小的彩色圆墩，使空间更具创新与自由。天花板以"蝴蝶"为元素，抽象化地设计出几个飞舞的蝴蝶造型，寓指路演成功后创业者破茧成蝶、展翅高飞。

开敞办公区域的中心设计了一个椭圆形的多用途空间，可以作为接待室、讨论室、会议室、休息室等使用。用不到顶的透明玻璃作为隔断，提示出这个空间的公共性及开放性，造型现代化的彩色家具很容易与彩色地毯融为一体，更增添了空间的穿透性。

2014年的巴西世界杯也给我们的设计带来了灵感。足球网作为隔断的一种材料及形式，也被我们运用到了不同区域的分隔中。这样既能使空间充满运动的气息，又不乏柔和的可触摸感。相信在世界杯期间，创业者会在这里感受到更多的激情。本案与大堂公共区域之间采用透明玻璃隔断，多彩的自由空间通过明亮柔和的灯光传递到大堂中庭，宣讲着自由与创新的每一个动人故事。

SOCIAL/WEB/ONLINE

Google Amsterdam

阿姆斯特丹谷歌办公室

Design Agency: Interior Designer: D/DOCK

Area: 3,000 m²

Photography: Alan Jensen

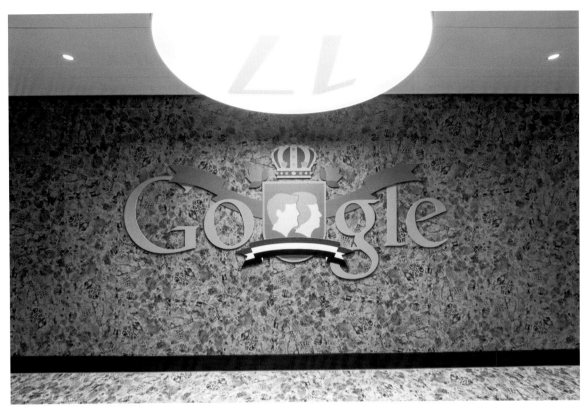

After an extensive refurbishment, Google Amsterdam reopened the doors of its quirky office in January 2014. Interior design studio D/DOCK was given the opportunity to turn this 3,000m² office space into an interactive landscape for the local Google team.

The garage where founders Larry Page and Sergey Brin started Google, was the inspiration for the interior concept. Quirky elements throughout the office illustrate this era from graffiti walls and cardboard box lights to the exposed ceilings and container wall in the 70-seater auditorium, also referred to as the Tech Talk.

Sustainability played a vital role in the restyling of the office. Existing meeting furniture, individual work places and parts of the micro-kitchens were offered a second life after refurbishment. D/DOCK adhered to Google's healthy material list by using non-toxic materials only and designed with a great focus on energy and water consumption.

Situated in the Southern part of Amsterdam, the office is considered as an intelligent landscape. Each floor features its own cave; the zone build around the core of the building where all general facilities such as meeting rooms, huddle rooms, video booths and micro kitchens are located. The neighborhoods with the individual work places surround this central cave and represent the various departments. Every work place is adjacent to the window affording all employees a 180° view overlooking the city while working. This layout makes the office space highly adaptable. It allows the Googlers to communicate and work together in a diverse environment while having personal space at the same time.

Each floor gives a nod to typically Dutch elements - whether it is the carrier cycle reception desk, the Stroopwafel ceiling panels, Gingerbread wall covering or the Delft Blue graphics in the restaurant. The re-used Febo snack wall that once served the Amsterdam cafeteria, now acts as a distribution point of computer accessories. For those wanting to spice up their meeting, the real life sixties caravan in the middle of the office is the solution. It is nowhere close to an ordinary meeting room with the comfy cushions and cozy vintage furniture.

According to D/DOCK's partner Coen van Dijck, the Amsterdam office is a feel good office. "It is a place that makes the employee perform better by offering a work environment that meets their needs", he explains. "Happiness, comfort, flexibility, relaxation, well balanced nutrition, exercise, daylight, fresh air and visual stimulation are some of the fundamentals that makes this office a healthy one", van Dijck continues. Google's Healthy Food program ensures well-balanced menus in the restaurant, seating up to 80 Googlers. The gym, meditation room and the desk bikes all offer opportunities to stay in shape, both physically and mentally.

  阿姆斯特丹谷歌办公室经过全面整修，终于在 2014 年 1 月展现了其奇异的面貌。D/DOCK 室内设计工作室有幸获得了这次机会，来将 3 000m² 的办公空间转变成当地谷歌团队的互动景观。

  谷歌创始人拉里·佩奇和塞吉·布林最初在车库中创立谷歌，这个车库就是此次室内设计的灵感源泉。办公室中的奇异元素诠释了这个时代：涂鸦墙、纸箱灯、未加装饰的天花板，以及"对话科技"礼堂（70 个座位）中的陈列墙。

  在办公室的改装过程中，可持续性扮演着重要的角色。办公室经过翻新之后，原来的会议家具、个人工作区域和小型厨房的部分空间就像获得了重生。D/DOCK 严格按照谷歌列出的健康材料单，采用无毒材料，注重节约能源和用水。

  办公室坐落在阿姆斯特丹南部，布局十分巧妙。每层楼都设有洞穴式的空间，每个分区都围绕着建筑的中心设置，通用设施齐全，如会议室、秘密会议室、视频台和微型厨房。中央空间周围设有单人工作位，代表了不同的部门。每个工作位都设在窗边，职员能够在工作间 180° 俯瞰城市风光。这种布局使空间灵活性更高，谷歌的职员能够在多样的环境中一起工作与交流，也可以拥有私人空间。

  不管是自行车接待台，还是荷兰松饼天花板，或是麦饼墙，甚至是餐厅中的代尔夫特蓝色图案，每层楼都与典型的荷兰元素相呼应。曾经用在阿姆斯特丹自助餐厅中的菲伯小吃墙，现在充当了电脑配件中的分配器。若是想让会议更加精彩，则可以选择办公室中央的六十年代大篷车，篷车内设有舒适的软垫和旧式家具，绝对能让你的小会议非同一般。

  根据 D/DOCK 合作人 Coen van Dijck 所说的，阿姆斯特丹谷歌办公室能给人一种美好的感觉。"在这样一个能满足职员需求的办公空间中，他们能够发挥得更好，"他继续解释道。"幸福、舒适、灵活、放松、锻炼、日光、营养均衡、新鲜空气和视觉冲击都是创造健康空间的基础，"van Dijck 说道。谷歌的健康食品项目确保了餐厅（可容纳多达 80 位职员）菜单所列食物营养均衡。健身房、静室和自行车办公桌都给职员提供了保持体形的机会，让身心更加健康。

Walmart Headquarter in São Paulo

圣保罗沃尔玛总部

Design Agency: Estudio Guto Requena

Architect: Guto Requena

Area: 6,400 m²

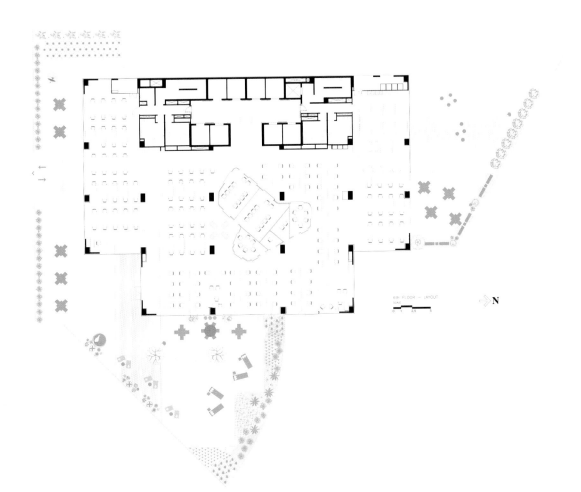

Pre-Design Research Methodology

The design for the Walmart Offices was derived from a research methodology developed by Estudio Guto Requena. Interviews and dynamic online exchanges with company employees were conducted to assess values, needs and expectations. 3 principal focal points emerged from this process: digital culture, the Walmart.com brand and brasilidade (Brazilian identity). This research also informed the choice of colors, materials, forms, programming and design concepts.

Concept Ual Framework

We applied these 3 focal points and their commonalities to an exploration of the building's prominent terrace and developed from this a guiding concept for the company's headquarters: the Urban Veranda. Design choices reference the Brazilian habit of engaging outdoor areas for social interaction and relaxation. Elements include beach chairs, the many large buildings with terraced facades, picnics (visible in the carpet patterning), the patios and balconies of Brazilian homes, and the rural habit of placing a chair in the street to enjoy the evening and chat with neighbors.

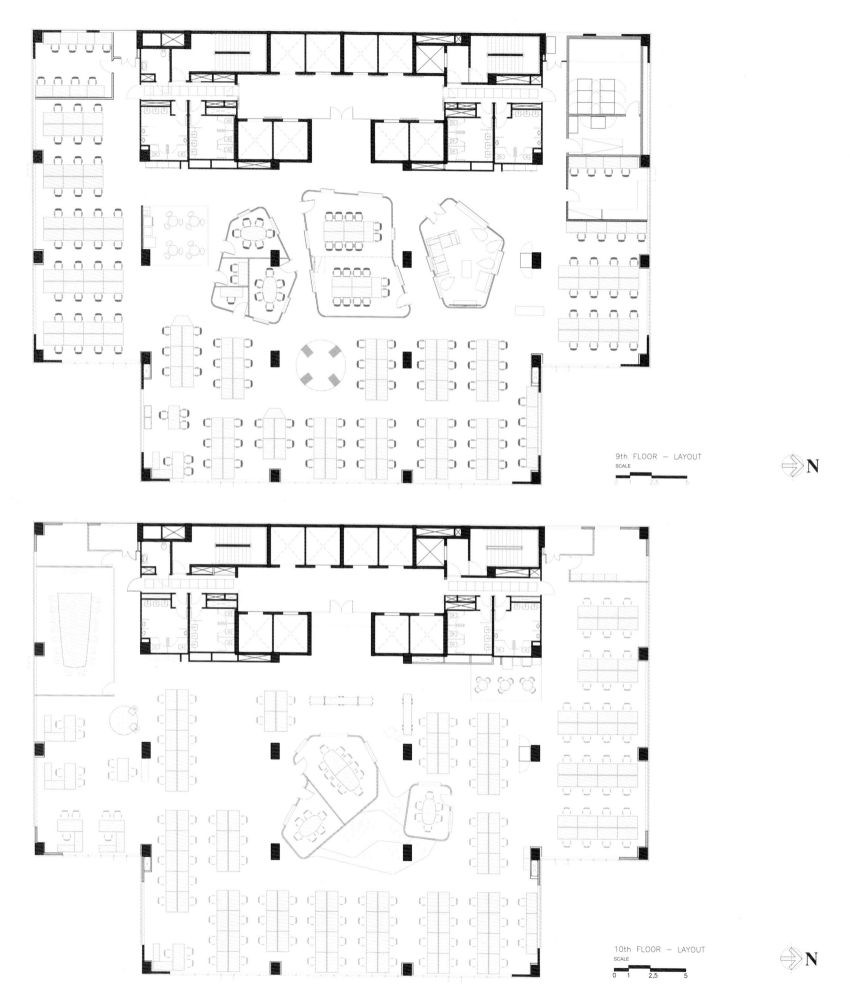

Program and Identity

The headquarters occupy 5 floors with over 1,000m² each. One of the challenges of this project was to bring a more human dimension to the work environment with spaces that are welcoming and comfortable, even pleasant and informal, while maintaining professionalism and practicality. Other challenges included a lean budget and a tight deadline.

To help locate and guide visitors and employees in this large area, we created a unique visual identity for each floor through centralized cocoons that develop organically between the pillars and break the rigidity of the orthogonal space. Each floor was designed with a predominant wood type. Pine, OSB, Eucalyptus and Masisa Zurich combine with a single color in various shades, all chosen from the official Walmart color palate of yellow, orange, blue and green. Different floors house individual departments, such as Business, Sales,

Human Resources or Finance, and also contain lounges and decompression environments, including games rooms, film screening areas, video games and a library. These areas are to encourage the exchange of ideas and interaction between employees from different departments.

Lighting

Workstations are located near windows to take advantage of daylight, and the lighting design prioritizes economy. In lounges and decompression areas indirect light is used in amber hues with decorative fixtures. Specifically created for this project is the hanging Gourd Lamp made from the fruit itself. Traditionally, these have been used in Brazil as containers, and also as resonators in musical instruments such as the chocalho, the berimbau and the maracá. Dried gourds were painted gray inside and arrayed on a wooden support, with colorful wiring left exposed.

设计前调查法

沃尔玛电子商务办公室设计方案是在古托·雷克纳建筑公司提出的一种研究方法的影响下得出的。通过设计前期与职员进行面谈和网上交流，对其价值观、需求和期望做了评估，从而得出了设计的3个要点：数字文化、沃尔玛品牌和巴西特性。

除此之外，调查还涉及色彩、材料、外形、结构和设计理念的选择问题。

基本概念

我们将这3个要点及其共性应用到了阳台的试探性设计中来，并从中探索出公司总部设计的指导性理念：城市阳台。设计参考了巴西人民参与户外社交和休闲的习

Furniture And Decoration

We prioritized the use of domestic furniture in both the offices and lounges, with signed pieces by the established Brazilians designers Maurício Arruda, Jader Almeida, Lina Bo Bardi, and Paulo Alves and Fernando Jaeger. We also included pieces that are parts of the popular Brazilian imagination, such as rocking chairs, beach chairs, porch chairs and picnic tables.

For the production of objects and decorative elements we used images of contemporary Brazilian Photographys, as well as maps, illustrations and Brazilian folk art. Skateboards and bikes reference the lifestyle s of younger employees.

Greenery

Throughout the office we emphasized the use of plants, and created a green belt that runs through the peripheral spaces and contributes to the identity and warmth of the work environment.

Terrace

The outdoor area was designed for both work and relaxation. Wood decking orders the environment, together with porch furniture, shaded areas, a space for yoga and a grandstand facing the facade that can host small events, concerts and film screenings. A mini-golf course was also speclally designed for the terrace.

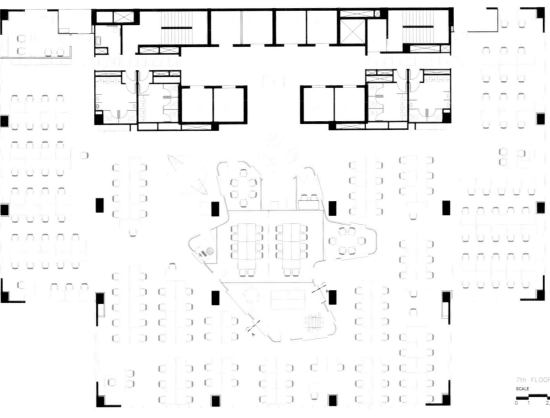

7th FLOOR — LAYOUT
SCALE
0 1 2,5 5

N

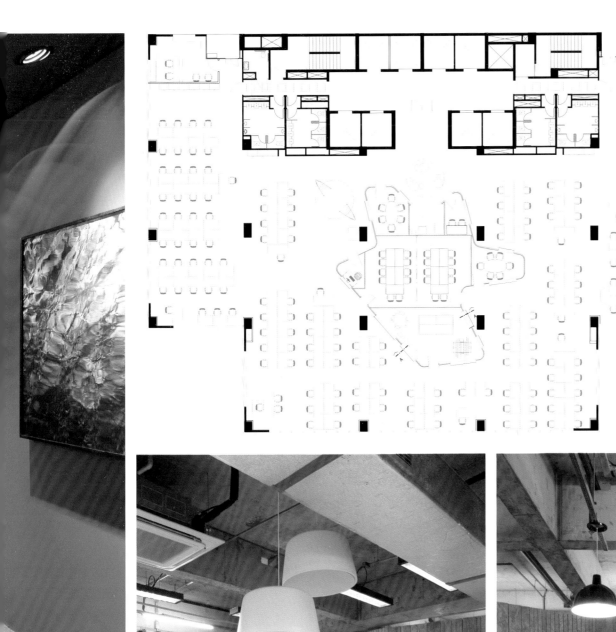

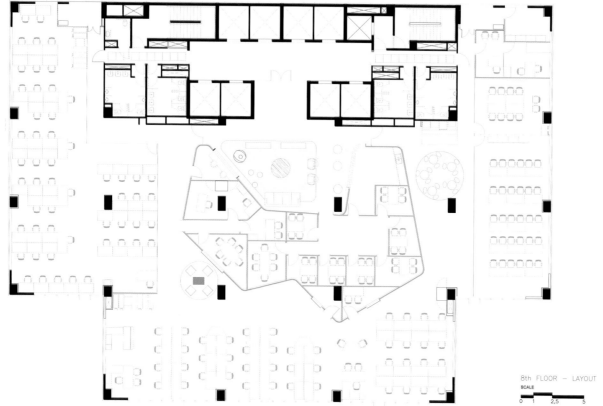

8th FLOOR – LAYOUT

惯。因此，所选元素包括沙滩椅、多数建筑采用的露台外观、野炊（地毯图案体现了这点）、巴西房屋的露台和阳台，还有把椅子搬到街上享受夜晚生活，与邻居闲谈的乡村习俗。

空间规划与特性创造

总部分为5层，每层楼面积超过1 000m²。该项目的难点是把人文环境带入工作环境中来，创造既舒适又受欢迎的工作空间，当然在保证空间的愉悦性和非正式性的同时，也保证空间的专业性和实用性。

蚕茧似的空间集中在柱子间，有序地排列着，既消除了矩形空间的刻板性，又营造出独特的视觉标识，能引导访客和职员在大型空间中移动。每层楼中都设计了一个醒目的木结构，如松树、刨花板、桉木以及苏黎世玛西沙木材，再用一种深浅不同的沃尔玛品牌色彩来点缀它，如黄色、橙色、蓝色和绿色。不同楼层设有不同的部门，比如业务部、销售部、人力资源部或财务部，同时，还设有休息区和缓解压力的空间，例如游戏室、电影放映室、电子游戏室和图书室。这些空间有利于促进职员之间和不同部门之间思想的传达和交流。

灯光设置

灯具设计优先考虑可经济性，因此工作区都设在窗边，有效利用了自然光线。在休息区和缓解压力区中采用了琥珀色装饰性间接照明灯具。而特别为本项目设计的"葫芦挂灯"是用水果做成的。在巴西，这些葫芦传统上被用作容器，还会在乐器中充当共鸣器，比如摇铃、拨铃波琴和响葫芦。干葫芦成排挂在木架上，内部被漆成灰色，而彩色的电线则暴露出来。

陈设和装饰

办公室和休息室都优先采用家用家具，还有知名的巴西设计师Maurício Arruda、Jader Almeida、Lina Bo Bardi、Paulo Alves、Fernando Jaeger设计并签名的家具。另外还有巴西流行的家具，比如摇椅、沙滩椅、走廊椅和野餐桌。

我们采用了当代巴西摄影中的意象，还有地图、插图和巴西民间艺术，来设计物品和装饰元素。滑板和自行车符合了年轻职员的生活方式。

绿色植物

植物被应用到了办公室的各个角落，形成了一条绿色带，贯穿了整个外围空间，使工作空间更加温馨，更有自身特色。

露台空间

露天空间是为工作和休闲设计的。这些空间以木甲板为主，并配有走廊家具、遮阳空间、瑜伽空间、观望台（面向可以举办小活动的舞台）、音乐会和电影屏。小型高尔夫球场也是专门定制的。

SOCIAL/WEB/ONLINE

Tencent Guangzhou Office

腾讯广州办公园区

Design Agency: M Moser Associates

Project Director: Wendy Leung, Joe Ho

Project Leader: Grace Hu

Concept Designer: Ramesh Subramaniam

Location: Guangzhou, China

Floor Area: 9, 914 m²

Photography: Vitus Lau @ M Moser Associates

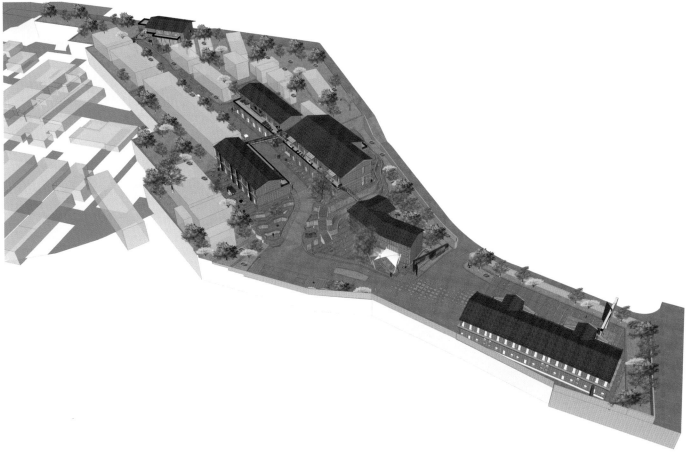

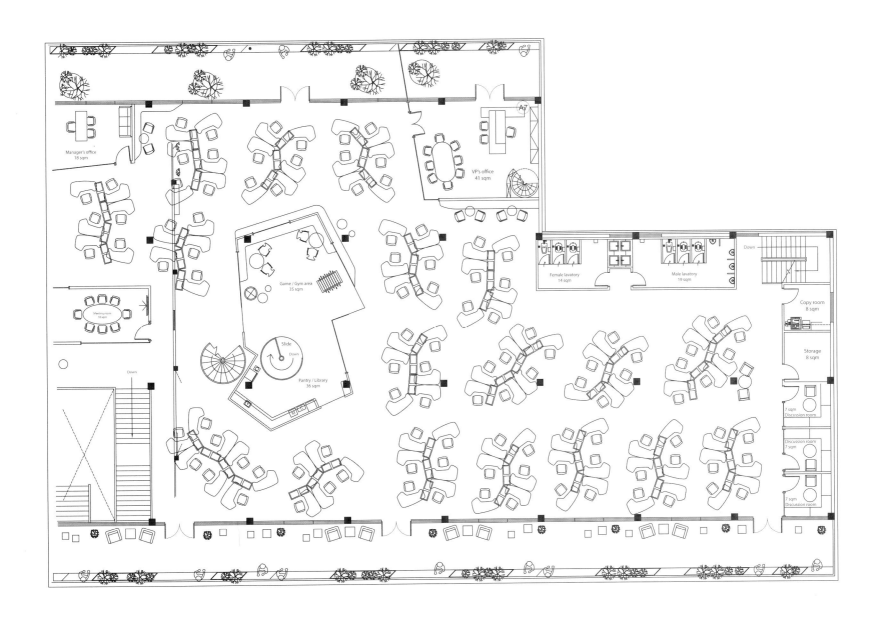

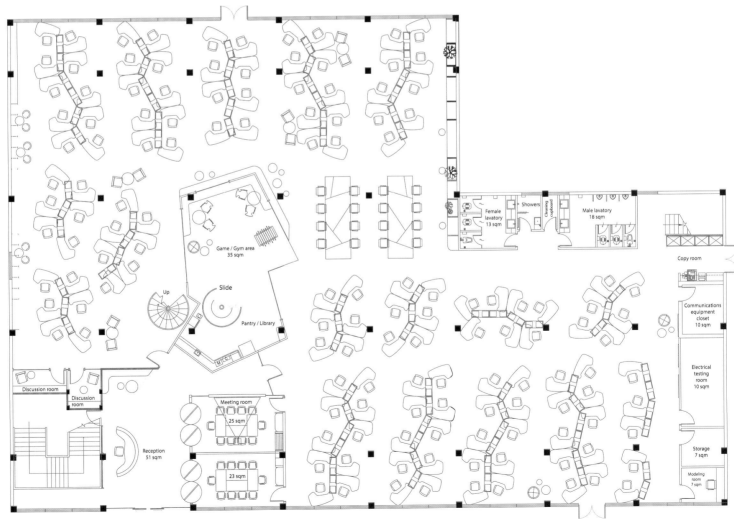

Old bricks, new purpose

This design was the result of a project to adaptively reuse 6 dilapidated factory buildings – part of a former textile industry site – into an inspiring, brand-defining workplace for Tencent, one of China's leading social media companies.

Externally the buildings were restored but little changed from their original appearance, preserving their industrial character and a part of local history. To form them into a single campus, they were linked together with open-sided bridges.

Inside, a key feature is the main entrance foyer with its full-height internal facade of meeting and work areas. Their irregular shapes, shifts of material and colour, and the insertion of natural foliage combine to suggest a stand of trees. This impression is furthered by the ceiling's 'cloud' contours and the natural light that floods the space.

An important aspect of the work areas are the mezzanine floors inserted into the buildings' existing shell. These maximise the ex-factory's vertical space, creating a "buildings within a buildings" composed of layers of open-plan work, meeting and social spaces.

Re-use of existing structures, the installation of efficient lighting systems, and careful sourcing of materials (including recycled, reclaimed and low-VOC materials) contribute to the high level of sustainability achieved by the project.

老砖头 新用途

该项目是翻修6栋废弃的工厂建筑，这些建筑属于原纺织工业区的一部分。现在设计成了中国领先社交媒体公司——腾讯的品牌性办公区，能给人带来无限的灵感。

建筑的外表经过了翻修，而且对外观也稍做了改变。作为当地历史的一部分，建筑的工业特性被保留下来。为了将这些建筑组成单一的园区，还设计了露天桥梁将它们连在一起。

内部重点设计了大厅，包括两倍高的会议区和工作区。它们有着不规则的形状，材料、色彩也变幻不定，还加入了天然的树叶，就像林立的树木。天花板上的"云朵"状轮廓和照射到室内的自然光线让这种感觉更加强烈。

工作区域的特色就是插入原有壳状空间的夹层。这种设计扩大了原空间的竖向空间，创造了一座"楼中楼"。夹层中设有会议、社交空间以及多层次的开放办公空间。

该项目重新利用了原建筑，安装了节能灯具，材料选用也十分细心（包括回收再利用的低挥发有机合成材料），从而成功打造了这栋可持续性极高的建筑。

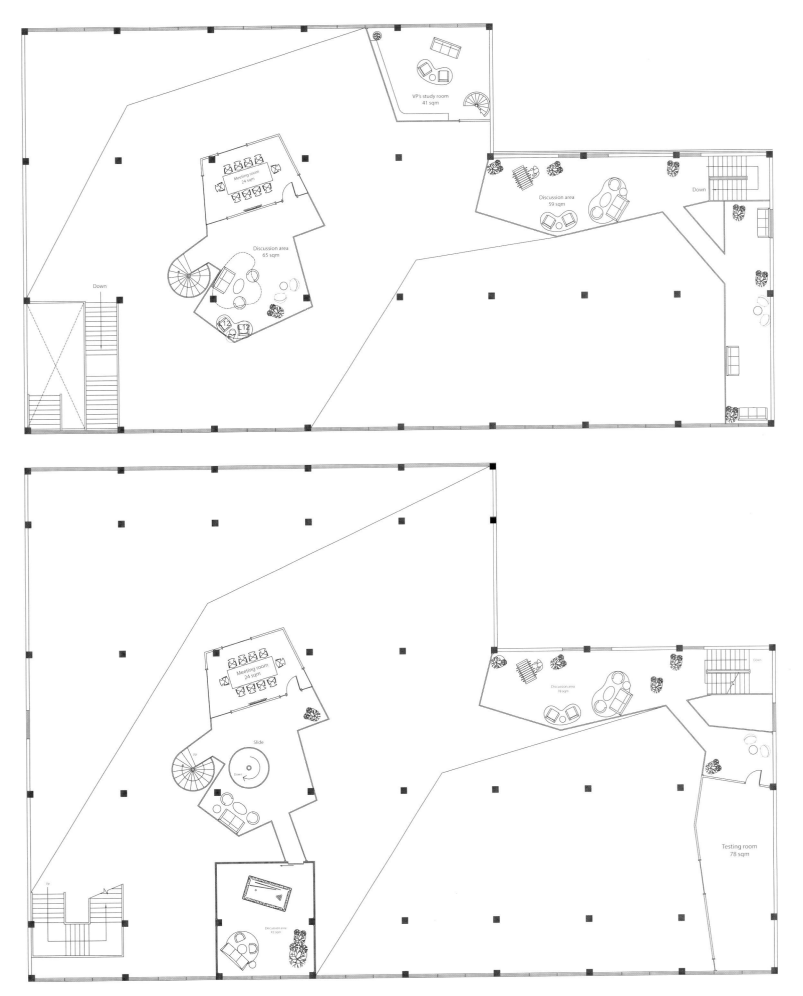

Contributors 设计师名录

3g office

3g office is an international company of Consultancy specialized in Workplace Innovation, Change Management and Facility Management, with large experience in corporative headquarters of big companies worldwide. We create tailored solutions where customer needs, best practices, and market trends are fit together to deliver a workplace where communication, productivity and employee satisfaction are improved. Our multidisciplinary teams address each project based on three pillars: Spaces, Technology and People, and are experts in Flexible Working and Flexible Office models.

AECOM

The AECOM Los Angeles Interior Design Studio is focused on creating high quality environments to work, collaborate, learn, relax and inspire. AECOM's interiors team is recognized for designing appropriate environments for clients across all market sectors and for bringing a new approach to traditional challenges. Our projects range from Fortune 100 company headquarters to boutique hotels to interiors for public agencies. In each of our projects, we act as stewards of our clients' aspirations and fiscal realities, working to capture opportunities and plan for the long term.

Led by Michelle Ives-Ratkovich—Design Principal and Director of Interior Design, Design Principal Vano Haritunians, FAIA and AECOM Americas Architecture Lead Ross Wimer, FAIA, the team's widely published design work has won prestigious awards including the Shaw Contract Group's design is...award and AIA awards. The DIRECTV Headquarters project was recently recognized as a 2014 finalist in Interior Design's Best of Year Awards (BoY Awards). The studio is currently at work on projects in throughout North America and Asia.

With nearly 100,000 employees – including architects, engineers, designers, planners, scientists and management and construction services professionals – serving clients in more than 150 countries around the world following the acquisition of URS, AECOM is a premier, fully integrated infrastructure and support services firm. AECOM is ranked as the #1 engineering design firm by revenue in Engineering News-Record magazine's annual industry rankings. The company is a leader in all of the key markets that it serves, including transportation, facilities, environmental, energy, oil and gas, water, high-rise buildings and government. AECOM provides a blend of global reach, local knowledge, innovation and technical excellence in delivering solutions that create, enhance and sustain the world's built, natural and social environments. A Fortune 500 company, AECOM companies, including URS Corporation and Hunt Construction Group, had revenue of approximately $19.5 billion during the 12 months ended Sept. 30, 2014.

ajc architects

Jill A. Jones founded ajc architects in Salt Lake City 23 years ago as a 100% woman-owned business, with pioneering determination and an unwavering commitment to value for her clients. Jill has crafted an architectural practice that is passionate about relationships, education, and the environment. While ajc's portfolio spans a diverse range of project types, we are most proud of those that provide solutions which improve our condition, as a whole.

In addition to our passion for improving the human condition, our team's expertise, technical skill, and design approach is always well-aligned with our clients core values. ajc employs the brightest and hardest working architects and designers and partners with the best engineers in the valley. We are dedicated to thoroughly, and quickly, exploring all necessary possibilities in order to ensure that we're working toward the best solution... and we won't settle until it's been achieved. ajc believes that environmentally responsible design is good design, and we incorporate sustainable strategies into all that we do - in our designs, and the way we operate and maintain our office. Our design work has afforded ajc unparalleled experience in implementing innovative, cost-effective, sustainable strategies into buildings and sites throughout the U.S. The long list of our projects designed/completed over the last decade which are LEED certified, or soon to be certified, by the U.S. Green Building Council is testament to our commitment, standards, and abilities.

ASTARTA

ASTARTA is the Russian manufacturer of implementing bold ideas in a complex arrangement of space for residential and commercial properties.

Belzberg Architects

Belzberg Architects has been located in the City of Santa Monica since its inception in 1997. The firm has earned over 48 national and local design awards including over 20 from the American Institute of Architects. Their work has been featured in over 200 publications throughout more than 25 countries including frequent features in notable periodicals such as Architectural Record, Interior Design, and the New York Times.

Belzberg Architects has been honored with the City of Los Angeles Cultural Affairs Committee Design Honor Award and the Urban Land Institute recognized their design work and their contribution to excellence in urban planning. The Chicago Athenaeum Museum of Architecture recognized several of their projects with multiple notable American Architecture Awards.

In 2007, the Association for Computer-Aided Design in Architecture, Canada featured their work at the Expanding Bodies Conference which focused on "digital sensing and interactive and responsive systems." Subsequent international exhibitions included "Material Skills: The Future of Intelligent Materials" and "Transforming the Built Environment", both of which "explored how digital fabrication can be integrated with existing design practices to improve the productivity and efficiency of construction and manufacturing." As a result, their research and built work was featured in and on the cover of the book Innovation in Building Design + Manufacturing for "spanning the historical gap between thought and hand, between idea and materiality."

In the past 10 years, Belzberg Architects has grown into a multi-faceted, professional design studio capable of producing large, and diverse sustainable building solutions and project types while maintaining a personal and intensive involvement with each project.

In 2013, the Museum of Contemporary Art in Los Angeles has chosen the work of Belzberg Architects to be included in their show "A New Sculpturalism: Contemporary Architecture from Southern California". This is the first extensive, scholarly examination of the radical forms that have become prolific in Southern California architecture during the past twenty-five years.

BORIS VOSKOBOYNIKOV

BORIS VOSKOBOYNIKOV is a well-known Russian interior designer and architect with more than 25 years' professional history.

In 1986 Boris graduated from Moscow State Industrial Art University, named after S.G Stroganov with specialization in "Interior Design". Since 1989, he has been a Member of the Artists' Union, Moscow branch.

Worked "solo" for several years and then in 2000 joined Nefaresearch team and the studio started its 'Golden period".

From 2003 to 2013 Boris headed the creative team and, being a partner, has leaded Nefaresearch to be one of the 20 best studios in Moscow.

In 2014 Boris Voskoboynikov have lounged VOX Architects – his new project, which accumulates the experience and logically extends its boundaries. VOX Architects specializes in modern architecture and interior design, creates brands and architectural concepts.

VOX Architects forms the project as a holistic environment that stimulates the customer to maximize his personal qualities and corporate aims. More than 100 projects are already completed.

Design - private and public spaces, café, boutique and restaurant interiors, kids clubs, offices, airports, metro station pavilions, residences, etc.

Full cycle of design works from the idea, working documentation and to whole project management and successful implementation, such as the equipment, lighting and furniture order and delivery management.

Cooperation with Russian, Dutch, German, Italian and British partners in the spheres of design, manufacture of equipment, materials, furniture and lighting.

Participation in professional exhibitions and competitions in Russia and around the world, repeatedly became the winner.

Creative credo:"Less but better"

CUBE architects

CUBE architects are an ambitious and enterprising office where strategic thinking is of great importance. We want to make a positive contribution to a sustainable society.

Our work includes sustainability advice, redevelopment and urban plans, but also interior design. We respond to current issues and are always looking for smart and high-quality opportunities for (re) development of buildings or (residential) concepts.

CUBE architects represent a substantive and comprehensive approach. A successful project uses all the features it has in itself, spatial, conceptual, programmatic, financial and functional. We develop in the beginning a good overall concept in which common principles be established.

We believe in the dynamics of the environment and the assignment. A design is the achievement of more than only m2's. It is a comprehensive strategy to create a more value, a smart and better space for users and future initiatives.

D/DOCK

D/DOCK is an Amsterdam based design studio of interiors and total concepts. The philosophy behind the studio goes beyond physical and aesthetic aspects alone. D/DOCK adopts a working method in which design, project management and consultancy are the ingredients for innovative business cases build around human requirements. The firm's team draws on an international network of experts, collaborating with respected professionals in the industry who share knowledge and passion for what they do, jointly seek solutions that stimulate ambition and education, and contribute to a sustainable future. D/DOCK develops and realizes innovative solutions for healthcare and educational institutions, retail, hospitality and work environments.

D/DOCK collaborated with national and international respected clients such as Google, Philips, B/S/H/, TomTom, VUmc Amsterdam and 20th Century Fox. The studio is currently working on the new work environment of Delta Lloyd, and the design of the 'Brede School' in Brunssum, operating in countries and regions as Austria, Germany, China, Taiwan, Scandinavia and the island of Aruba.

dan pearlman

dan pearlman is a strategic creative agency.

Our goal is to anchor brands and experiences in the hearts and souls of the people. To deliver this we combine an extensive, strategic competence with high-level creativity in implementation.

In action we take a 360° approach. We cover the entire spectrum, from strategic positioning and developing creative ideas for brands to implementing them in design and media, retail and brand experiences as well as architecture for zoos, hospitality and leisure environments. We offer all the necessary skills under one single roof. Our people are highly talented. They are passionate about their own area of specialisation. And they are able to see the big picture, and beyond.

DESIGN BLITZ

Design Blitz was founded in 2009 by Partners Melissa and Seth Hanley. As a full service architecture and interior design firm, we provide the complete range of architectural services required to take a project from programming through construction. Though our focus is the built environment, we are committed to total design solutions – balancing buildings, branding, and experiences.

Design Blitz leaves many of the inefficient conventions of traditional practice behind by systematizing delivery, removing unnecessary management layers and leveraging technology to reduce errors, paper, and shorten the feedback loop wherever possible. By removing the barriers to open communication and flattening the delivery process, we deliver high quality projects faster.

In short: We Blitz It.

DSP Design Associates

DSP Design Associates was established in 1988 as a Pan India multi-disciplinary and integrated design consultancy firm, celebrating 25 years of design innovation, deep understanding, and relentlessly high standards. This uniquely talented firm has claimed the length and breadth of India as its stage, building a formidable reputation for award-winning designs, and an adherence and expertise in 'green' sensibility.

This promise of excellence has drawn a marquee clientele that includes multinational organisations, leading developers of the country, world class hotel franchises, leading business houses and many more.

DSP's mission is to bring an unusual degree of quality, reliability, and refinement to the table and provide a personalized & total quality effort focusing our ability to deliver exceptional service to exceed Clients expectations.

Estudio Guto Requena

Estudio Guto Requena reflects about memory, digital culture and poetic narratives in all design scales. Guto, 34 years old, was born in Sorocaba, countryside of Sao Paulo State. He graduated as Architect and Urban Planner in 2003 at USP - University of São Paulo. During nine years he was a researcher at NOMADS.USP - Center for Interactive Living Studies of the University of São Paulo. In 2007 he got his Master degree at the same University with the dissertation, Hybrid Habitation: Interactivity and Experience in the Cyberculture Era.

He won awards and had lectured and exhibited in several countries. He was a professor at Panamericana - School of Arts and Design and at IED - Istituto Europeo di Design - at both graduation and master levels. Guto had lectured on 70 workshops all over the country, and received the Young Brazilian Awards recognition. In 2012 Guto was selected by Google to develop the project for their Brazilian headquarter. And in 2013 Walmart selected him to design their headquarter.

Since 2012 Guto has a column at newspaper Folha de São Paulo where he writes about design, architecture and urbanism and collaborates writing for many magazines. In 2011 Guto created, wrote and hosted the TV show Nos Trinques, for Brazilian TV Globo channel GNT and developed design web series for the same channel, recorded in Milan, Paris, Amsterdam and London.

Fokkema & Partners Architecten

Fokkema & Partners Architecten is often asked to describe the firm's chief characteristics. Just naming a single style doesn't serve to answer this question, since the projects vary so widely. Still, since the start of the firm in 1995 all projects have had one thing in common: a drive for quality that exceeds their clients' expectations. It's our ambition to get the most out of a design question on their clients' behalf - a thorough process that starts in understanding clients needs through carefully and methodically trying to come to the core of the matter.

Fokkema & Partners Architecten encourage clients to come up with considerations and critical questions during the concept development phase. They fee that concept development really is one of the most important phases of the design process, a stage with a clear moment, when the true nature of the solution is determined. It is a phase when out of the box thinking, sharp analysis and down to earth practicality all come together in their effort to constantly raise the bar and stay away from the obvious.

During the next phase of design development we keep utilizing their creative capacities to find solutions which stay away from the obvious but still remain practical. At the end of the day they like to be inspired and have fun, but most importantly they want our client to be proud of their common result.

Giant Leap

Whatever your business, you want to create spaces that inspire people. Giant Leap is passionate about designing superior interiors that take you on a journey the minute you step inside. We share a passion for excellence in design, delivery, and finish, always using the finest materials, technology, logistics and people. We create better places to work, rest, play, be original, be comfortable, and above all, be inspired. This is the heart of Giant Leap.

HEYLIGERS design+projects

HEYLIGERS design+projects is an agency for interior design and architecture with approximately 15 to 25 employees in two offices in Amsterdam and Utrecht. In its 25- year existence, the agency has established itself as a specialist in the design and management of complex and innovative interior design commissions for the business market t in the Netherlands and abroad. The research and implementation of new workplace concepts is often part of the scope. In all projects we aim towards an end result that holds the maximum achievable sustainability qualification. HEYLIGERS design+projects has broad experience in achieving LEED, BREEAM and Greencalc + certification.

IND Architects Studio

IND Architects Studio is featured by a particular attention to details. They believe that it is a detail that shows the quality of architecture.

Their team consists of experienced architects who develop the projects starting with a sketch and following it up to complete implementation of intended ideas.

They are united by true passion and commitment.

Since the studio was founded in 2008, they have been dealing with design of apartment and public buildings, town houses and interiors, office premises, hotels, business-centers and restaurants. In these latter days, they are engaged in development of landscape and urban design areas and widely participate in various competitions.

Their strengths: individual attention to every client, effective system of business processes, flexible and quick decision-making, complex approach to architecture and interior, thorough researches and quality analysis at all project phases, quick adaptability to new market requirements.

They provide their clients with comprehensive design documents prepared in accordance to high standards.

Their clients are comprised of those who generate the demand for high-quality architecture: individuals, state officials, businessmen, development corporations.

You can find their works in the most picturesque places of Russia, Spain, Montenegro and Kazakhstan. They are keen to expand this list and take part in up-market international contests.

Ippolito Fleitz Group

Ippolito Fleitz Group is a multidisciplinary, internationally operating design studio based in Stuttgart. Currently, Ippolito Fleitz Group presents itself as a creative unit of 45 designers, covering a wide field of design, from strategy to architecture, interiors, products, graphics and landscape architecture, each contributing specific skills to the alternating, project-oriented team formations. Their projects have won over 200 renowned international and national awards.

Jannina Cabal & Arquitectos

Jannina Cabal Arquitectos Studio was founded in 2003 after 4 years of work and experience in construction companies and architectural consulting firms. Since the beginning, the acceptance received by private clients and real estate developers was outstanding, generating a positive commercial growth to the studio. JCA presented a very clear and defined design style, recognized by clients who arrived seeking for innovative proposals. This happened within a competitive and complicated field that thanks to the work of young architects full of creativity and energy. Therefore, they had to deal with many new challenges that demanded a lot of effort, dedication and progressive work.

Today, the studio has a team of 10 architects, specialized in several areas, in addition to the group of consulting engineers that collaborate with JCA.

In the present the firm is involved in a variety of successful residential, urban and commercial projects within Ecuador.

Johnson Chou Inc.

Since 1999, Johnson Chou Inc. has developed into an interdisciplinary design practice encompassing architectural and industrial design, furniture and interiors, graphic identity and corporate communications - a body of work characterized by conceptual explorations of narrative, transformation and multiplicity.

While the search for the elemental is the defining aspect of their work, elements of drama and engagement – on intellectual, emotional and physiological levels of experience - exemplify the firm's projects.

The creation of a narrative forms the conceptual point of departure for all of the firm's work and is essentially a story inspired by the client. Be it about the quirks of the client, an image they wish to project, or even delved from a literary source, it is an interpretation of the client's image, culture or brand – what the firm likes to describe as "narratives of inhabitation".

The firm's award-winning and internationally recognized projects include advertising offices for Grip Limited and Zulu Alpha Kilo, Head Offices for Red Bull Canada and private residences in 10 Bellair and the Candy Factory condominiums. Other projects include the Museum of Canadian Contemporary Art (MoCCA), TNT and 119 Corbo clothing boutiques, and Blowfish Restaurants.

LAB5 architects

LAB5 architects is a Budapest (EU) based design studio established in 2007, covering full services for urban, architectural, and interior design. The company provides characteristic solutions, where the iconic look-out is combined with pragmatical functional layout, usually on European and Asian sites.

Li Yiming

Li Yiming First grade registered architect
Senior Engineer
Design Director, Beijing Qingshi Architectural Design and Consulting Co., Ltd.

He graduated from Northern Jiaotong University in 1998, majoring in Civil and Structural Engineering. Since graduation, he has been engaged in architectural solutions and managing design projects in large state-owned design institution. He has worked on over 20 large projects one and after another, covering urban mixed-use project, commercial center, five-star hotel and large-scale science and technology park, etc.

Representative works include Zhongguancun Dongsheng Science and Technology Park, Taishun Cultural and Creative Industrial Park, Beijing Four Seasons Creative Park, Beijing Union University Zhongguancun City of Science, Art Center of Capital Norman University, Xi'an Hua Qing Palace Yutang Spa & Resort Hotel, Ningxia Shapotou State Guest House, Yunnan Anxia Riverside Resort, etc.

Lv Xiang

Lv Xiang Senior interior architect
Technical Director, Beijing Qingshi Architectural Design and Consulting Co., Ltd.
He graduated from Shandong Institute of Urban Construction in 1997. Since graduation, he has been engaged in interior design and project management, including large office, financial institution, high-end hotel, restaurant and suchlike hundreds of projects.

Representative projects include Bank of China Financial Center, Chengtong Group Headquarters, Management Committee Center Building of Chengdu Hi-tech District, Beijing Shijingshan District Court, Beijing Century Technology and Trade Building, Beijing Dong An Men Hotel, Beijing Gloria Plaza Hotel Dongsheng, Dalian Bangchui Island Resort Hotel, Sichuan Zigong Grand Hotel, etc.

MoreySmith

MoreySmith is an architectural design practice based in London. Founded in 1993 by Linda Morey Smith, the practice has an extensive portfolio of commercial, workplace, development, leisure and residential projects. The company mission is to deliver beautifully designed projects which reflect and embody the client, their brand ethos and needs. Projects are undertaken by employing a strategic approach underpinned with design-led and detail-focused thinking.

The practice offers architectural design, interior and exterior refurbishment, interior design, branding and property services. MoreySmith works with both occupiers and developers. MoreySmith is passionate about designing environments where people love to work.

M Moser Associates

M Moser Associates is a global firm specialising in the architecture, engineering, interior design, and delivery of corporate workplaces. Clients range from multinational companies and financial institutions to privately-owned businesses.

A key to how M Moser fulfils clients' business needs is the Integrated Project Delivery (IPD) approach. This enables project teams and services to be precisely focused on the unique demands of each client and project. IPD also enables spaces and buildings to be developed and delivered as truly holistic solutions to clients' specific needs.

M Moser's 700+ staff includes experienced planners, interior designers, architects, engineers, IT specialists and construction professionals. Founded in 1981 - and in 15 locations worldwide - the company currently operates in Bangalore, Chengdu, Beijing, Delhi, Guangzhou, Hong Kong, Kuala Lumpur, London, Mumbai, New York, San Francisco, Shanghai, Shenzhen, Singapore and Taipei.

NL Architects

NL Architects is an Amsterdam based office. The three principals, Pieter Bannenberg, Walter van Dijk and Kamiel Klaasse, officially opened practice in January 1997, but had shared workspace already since the early nineties. All were educated at the Technical University in Delft.

NL Architects aspires to catalyze urban life. The office is on a constant hunt to find alternatives for the way we live and work. How can we intensify human interaction?

Some of their projects include Parkhouse/Carstadt (an attempt to integrate auto-mobility and architecture), WOS 8 (a seamless Heat Transfer Station) and the Mandarina Duck Store in Paris. The BasketBar (a grand café with basketball court on the roof) and A8ernA, the redevelopment of the space under an elevated highway, have become emblematic contributions to contemporary culture.

OSO Architecture

OSO Architecture is an Istanbul-based architectural studio established by Okan Bayık, Serhan Bayık and Ozan Bayık in 2007. A strong emphasis is given to the critical design process within the studio; we resist predetermining architectural solutions to a client's brief prior to a thorough investigation of each project's unique situation.

Our criteria for design is to pay close attention to contemporary design methods, new materials and the economic considerations of the client. We believe that a good project must combine all these areas. Moreover, the most important thing for us is to look at a project from a new perspective.

In order to achieve these aims, OSO Architecture has brought together three different professions; architecture, interior design and civil engineering.

OSO Architecture specialises in architecture, interior design and project management. Our aim is to create examples which are not only unique, but also combining efficiency, economy and design.

PENSON

PENSON focuses on providing talented architecture, interior design, structural, civil, mechanical & electrical engineering consultancy services, specialising in all sectors of buildings & uses.

They are unique because our chartered architects, designers & engineers sit together, to simplify communication whilst improving coordination & efficiency.

We grow by working hard to create award winning solutions, whilst paying close attention to every project's commercial & deliverable needs.

Their quietly competent approach means that we act repeatedly for the world's biggest & smallest organisations in both public & private sectors, with the occasional film, pop & sports star.

Their largest project is in Australia, where we are creating a new media & leisure village at $1.2billion. Their smallest project is a bespoke gallery for The Arts Council at a valued £290k, which is featured in the New York Times, ICON, FX, BUILDING, Mix & many other highly respected journals. It was also featured on this years Stirling Prize show on BBC2.

Their team unites London's best young talent leading the field of design & sustainability. They help to judge international design competitions, speak at conferences, teach architecture & write articles for some of the world's leading design journals. We also host frequent events for numerous charities for fun & enhanced team building.

We have won many international design awards & solve problems.

It just so happens that we draw too.

Perkins Eastman

Perkins Eastman is among the top design and architecture firms in the world. With 900 employees in 14 locations around the globe, Perkins Eastman practices at every scale of the built environment. From niche buildings to complex projects that enrich whole communities, the firm's portfolio reflects a dedication to progressive and inventive design that enhances the quality of the human experience. The firm's portfolio includes high-end residential, commercial, hotels, retail, office buildings, and corporate interiors, to schools, hospitals, museums, senior living, and public sector facilities. Perkins Eastman provides award-winning design through its offices in North America (New York, NY; Boston, MA; Charlotte, NC; Chicago, IL; Los Angeles, CA; Pittsburgh, PA; San Francisco, CA; Stamford, CT; Toronto, Canada; and Washington, DC); South America (Guayaquil, Ecuador); North Africa and Middle East (Dubai, UAE); and Asia (Mumbai, India, and Shanghai, China).

plajer & franz studio

plajer & franz studio is an internationally active agency for intelligent brand architecture providing design concepts, interior design, retail design and architecture for retail environments, department stores, office and hospitality spaces.

Our services range from corporate identity and concept design to roll-out and graphic support. This multi-specialist approach ensures freshness of vision and a broad-minded attitude.

pS Arkitektur

pS Arkitektur works with projects ranging from commercial interiors to private houses and urban planning. Our vision is to create unique buildings, interiors and environments that make an emotional and visual impression. Our motto is "architecture that makes a difference".

We challenge the obvious solutions and aesthetics in favor of creating something unique, derived from the company's identity and our client's dreams. Our office interiors and refurbishments enhance the clients brand identity and business opportunities.

Architecture and design is a means of market positioning, creating far reaching values for our clients. We define the goals and demands so that the design will be an efficient tool and identity bearer. We bring forward the possibilities, visualize the invisible and suggest changes. We believe that inspiring environments generate positive energy!

In 2011 we received first prize for Outstanding Design of the Year at the 9th Modern Decoration International Media Award in Hong Kong. Silver in the Swedish award Swedens Prettiest Office, as well as nominations for best office design in Leaf Awards in London and Inside Festival in Barcelona.

Resonate Interiors

Resonate Interiors is born out of Pernille Stafford's desire to create memorable environments, special spaces that engage with the occupant at many levels, with the primary focus on commercial interiors, offices, hospitality, retail and education environments. Resonate offers innovative crafting of space in three dimensions, the juxtaposition of materials to the manipulation of light and shade.

Setter Architects

Setter Architects was established in 1985 by Michael Setter, a graduate of the School for Architecture and Town Planning at the Technion Institute of Technology, Haifa, Israel. Setter Architects is considered one of the leading interior design firms in Israel, servicing corporations as well as various private and public organizations. The firm specializes in modern design with an emphasis on integrating international style with the client's unique organizational culture.

Setter Architects is comprised of elite architects and interior designers whose relationships with professionals in a wide range of related disciplines, enables them to create the best possible designs and business image. The firm's expertise at combining creative design with the needs of the client, results in projects that lend prestige to the client's business or organization.

Setter Architects' clients include some of the largest and most influential companies in Israel and in the world, including corporations in the hi-tech, pharmaceuticals, finance, and the automotive industries, as well as governmental offices and others. The firm's experience enables them to provide a comprehensive solution for every client request. Whether managing large-scale design projects for corporate campuses or shopping centers, or working with mid-size companies looking to upgrade their image, or even with private residences, Setter Architect's designs ensure a natural integration with the surrounding public areas.

Sergey Estrin Architectural Studio

Sergey Estrin graduated from Moscow Architectural Institute in 1985. In just five years he entered the top ten young architects of the Soviet Union, having won the All-Soviet Union tender for work in the capitalist country Ireland.

Experience of the foreign practical training abroad was not in vain, when Sergey returned very soon he got a prestigious position in the foreign company Global Resource Group. Having worked for such well-known companies as DTZ, Capital Group, etc., where he was in charge of design departments, Sergey established his own architectural studio.

With more than twenty five years of work experience in architecture Sergey Estrin became a high-profile person not only in the architectural space but also in the real estate market. Sergey was at the very beginning of formation of the Russian way of architectural thinking which started at the period of changes in political, cultural and social systems.

Sergey Estrin personally takes part in each project of the company and is the author of the most ideas. The main principle which appears in every work is the space and shape, full of emotion, light, color, combined with the sophisticated chic and comfort. Creative motto: "There are no desperate situations and hopeless clients".

Snook Architects

Snook Architects is an award-winning architectural practice based in Liverpool. Founded in 2005, they have completed projects across a variety of sectors from £100k to £5m. With a pragmatic approach, each project responds to its specific context rather than to a pre determined style, to create work with honesty and longevity.

SPACE | Juan Carlos Baumgartner

Juan Carlos Baumgartner studied for his degree at the Universidad Nacional Autónoma de México [UNAM, National Autonomous University of Mexico] Faculty of Architecture.

In April 2014, Mr. Juan Carlos Baumgartner obtained the Grants Hermes & Iris Prize for Leadership and Communication in Interior Design from the CIDI Iberoamerican Council of Interior Designers.

Space, the architectural firm led by Mr. Juan Carlos Baumgartner has received national and international recognition, including the Property Awards 2012 in London and the Design Award in Milan with the BASF project; in Italy it obtained the prize in the silver category with the Google project, bridgestone in the bronze category and the Design Award was obtained with the Volaris project.

Space has been ranked by Obras magazine for the third time as one of the 5 most important architectural firms in the country; in 2013 it was considered the most sustainable firm and in 2014 the most innovative firm in Mexico.

One of the most recent projects of SPACE is the EFIZIA high performance sustainable tower, which holds a LEED Gold Level pre-certification, being the first project to obtain this certification in the country.

Studio 13

As Studio 13, we create modern and dynamic living areas bringing personal and corporate tastes together with functionality and esthetic. With the creative architectural perspective, Studio 13's professional team transforms spaces into living areas with timeless designs.

Studio 13 leads with its dynamic projects that bring together its expertise and creativity for global brands; one of the best examples among our many projects is Unilever's Algida factory. The factory is located in Konya/ Turkey and it is Unilever's globally first factory with environment friendly LEED certificate. We focus not only on our corporate customers' but also our individual customers' needs by bringing together chicness and esthetic that we design for all housing and other private projects.

Talent Garden

Talent Garden is the first international network of Co-Working Campuses with a focus on digital. The first seat had been opened in 2011 Brescia, Italy. Talent Garden is a co-working federation with a model alike to franchising which allows creating new campuses all around the World.

It is a space for tech/digital/creative entrepreneurs, freelancers, startups, companies - all the digital ecosystem members to meet, work, learn and collaborate with each other. TAG also runs events, business development sessions, and educational activities and helps traditional businesses to innovate.

During the first years TAG has been opened in other Italian cities: Bergamo, Cosenza, Genova, Milano, Torino, Pisa e Padova. Current openings are happening in Luxembourg, Switzerland, Sri Lanka and NYC.

The Bold Collective

The Bold Collective was formed in September 2011 when the two principals of the practice decided to take the leap. The company has since grown to a team of five and we have delivered a number of corporate and retail projects. We are a multidisciplinary studio offering both interior and graphic design services and have quickly carved out a niche for ourselves within the media sector where we have created a series of dynamic workplaces for cutting edge media agencies. We are proud to have already worked with large organizations such as ABC and WPP. We believe that our approachable and friendly culture and our passion to create highly creative interiors, makes the process of working with The Bold Collective an enjoyable experience for both staff and clients. Our studio is design -curious and we are constantly sharing ideas from widespread sources, which helps to inform our work and ensures that our projects are diverse and fresh. We are keen to provide interiors that go beyond meeting the functional requirements of the brief to inspire, challenge and provoke.

ZGF Architects LLP

ZGF Architects LLP is a design firm with a focus on architecture, interior design and urban design with a mission to strive for design excellence, stewardship of our natural and built environment, and exceptional client service. Since being founded in Portland, Oregon more than 70 years ago, ZGF has grown to include over 500 employees and additional offices in Los Angeles, Seattle, Washington DC, New York, and Vancouver, British Columbia.

ZGF's design philosophy is centered on the premise that design excellence should be reflected in each and every aspect of a building—its fit with the community, its function and relationship to its users, its building systems, and its cost. The firm's design portfolio is intentionally diverse, ranging from transportation terminals, commercial office and mixed-use developments, to corporate campuses, healthcare and research buildings, academic facilities, libraries, and museums.

vib architecture

vib architecture was created by Bettina BALLUS from Germany and Franck VIALET from France to develop Vialet architecture's projects in a creative and sustainable way. Since 1996, we have designed and constructed laboratories, offices, housing, educational and cultural buildings across France in places like Paris, Toulouse, Caen, Bordeaux and Strasbourg. Our team of 20 architects is based in Paris and Bordeaux. For every project, we put together a specific team of partners and engineers, united around our design. Our strong and cosmopolite experience has led us to try to develop worldwide.

Our office has won the prestigious Price for the «Première Oeuvre» for our Research Center built in Caen.

"Our projects are always optimized and functionnal, but they also propose a poetic and conceptual dimension that guide the way we assemble space, matter ans light. This makes each of our projects unique."

[ARTPOWER]

The Renzo Piano Building Workshop

The Renzo Piano Building Workshop (RPBW) was established in 1981 by Renzo Piano with offices in Genoa, Italy and Paris, France. The Workshop has since expanded and now also operates from New York.

The Workshop permanently employs about 100 architects together with a further 30 support staff including CAD operators, model makers, archivers, administrative and secretarial staff.

Our architects are of many nationalities and each is selected for his or her experience, enthusiasm and calibre.

Our staff has a wide experience of working in multi-disciplinary teams on building projects in France, Italy and abroad.

As architects, we usually provide full architectural design services and consultancy services during the construction phase. Our design skills also include interior design services, town planning and urban design services, landscape design services and exhibition design services.

The firm is licensed to practice architecture in France. Renzo Piano personally is registered as an architect in France and Italy.

The Workshop has successfully undertaken and completed projects around the world.

Currently, the main projects in progress are: the Harvard museums in Cambridge, Massachusetts; the new Whitney Museum and Columbia University's Manhattanville development project in New York City, the Stavros Niarchos Cultural Center in Athens; the Valletta City Gate in Malta, the Paris Courthouse in Paris and the Academy Museum of Motion Pictures, Los Angeles.

Major projects already completed by RPBW include: the Centre Georges Pompidou in Paris; the Beyeler Foundation Museum in Basel; the Rome Auditorium; the reconstruction of the Potsdamer Platz area in Berlin; the Menil Collection in Houston, Texas; the New York Times Building in New York; the Kansaï International Airport Terminal Building in Osaka; the Kanak Cultural Center in Nouméa, New Caledonia; the California Academy of Sciences in San Francisco; the Chicago Art Institute expansion in Chicago, Illinois; the London Bridge Tower in London; the Kimbell Art Museum expansion and others throughout the world.

Exhibitions of Renzo Piano and the Building Workshop's works have been held in many cities worldwide, including major retrospectives exhibition in 2000 at the Centre Pompidou in Paris and at the Neue National Galerie in Berlin, in 2007 at the Milan Triennale and in 2014 at Palazzo della Ragione in Padua.

Acknowledgements

We would like to thank all the designers and companies who made significant contributions to the compilation of this book. Without them, this project would not have been possible. We would also like to thank many others whose names did not appear on the credits, but made specific input and support for the project from beginning to end.

Future Editions

If you would like to contribute to the next edition of Artpower, please email us your details to: artpower@artpower.com.cn